Know-It-All

Chemistry

Know-It-All

Chemistry

**The 50 Most Elemental Concepts in Chemistry,
Each Explained in Under a Minute >**

Editor **Nivaldo Tro**

Contributors

Jeff C. Bryan

Stephen Contakes

Glen E. Rodgers

Ali O. Sezer

James Tour

Nivaldo Tro

John B. Vincent

—**WELLFLEET**—
P R E S S

Quarto is the authority on a wide range of topics.

Quarto educates, entertains and enriches the lives of our readers – enthusiasts and lovers of hands-on living.

www.QuartoKnows.com

© 2017 Quarto Publishing plc

First published in the United States of America in 2017 by Wellfleet Press, a member of Quarto Publishing Group USA Inc.
142 West 36th Street, 4th Floor
New York, New York 10018
www.QuartoKnows.com

10 9 8 7 6 5 4 3 2 1

ISBN: 978-1-57715-151-7

This book was conceived, designed, and produced by

Ivy Press
An imprint of The Quarto Group
The Old Brewery, 6 Blundell Street
London N7 9BH, United Kingdom
T (0)20 7700 6700 F (0)20 7700 8066

Publisher Susan Kelly
Creative Director Michael Whitehead
Editorial Director Tom Kitch
Commissioning Editor Stephanie Evans
Project Editors Jamie Pumfrey,
  Joanna Bentley
Designer Ginny Zeal
Illustrator Steve Rawlings

Printed in China

CONTENTS

INTRODUCTION
Nivaldo Tro

The core idea of chemistry is that *the whole can be explained by its parts*. The properties of matter can be explained by the bits that compose it. Understand the bits, and you understand the whole. Philosophers call this idea reductionism. Reductionism has not always been popular in the history of thought, nor is it clear that it is universally true. But the stunning and ongoing success of chemistry in explaining the behavior of matter—even living matter—suggests that, at a minimum, reductionism is a powerful and useful idea.

The "bits" in chemistry are atoms, ions, and molecules. Although the idea that matter has fundamental "bits" is quite old—it was first suggested more than two thousand years ago—its broad acceptance is quite recent, and occurred only about two hundred years ago. Before that time, most thinkers thought that matter was continuous, that it had no smallest bits. The advent of the scientific revolution in the sixteenth century led thinkers to correlate their ideas about nature more carefully with empirical measurements. Since empirical measurements supported the particulate model, the continuous model was discarded.

Chemistry helps us understand that we—and all things around us—are made up of particles.

Once the particulate model was accepted in the 1800s, progress came relatively quickly. Scientists began figuring out the structure of the basic particles that compose matter, and by the early-to-mid twentieth century, chemists had good models that explained how atoms bond together to form molecules, and how the structures of atoms and molecules affect the properties of the substances they compose. In fact, throughout chemistry, the relationship between structure and properties is a key unifying theme.

A second unifying theme of chemistry is the progression from simple to complex. It turns out that, in nature, when you put together simple particles in slightly different ways, you can get vast complexity. Just as the twenty-six letters of our alphabet can be combined in different ways to compose many words, and just as you can combine those words in many ways to form an even larger number of complex ideas, so the ninety-one elements that compose matter can be combined to form many compounds, and those compounds can be combined to form an even larger number of complex substances, including all living things.

Graphene is a new, carbon-based material that is just one atom thick but is stronger than steel.

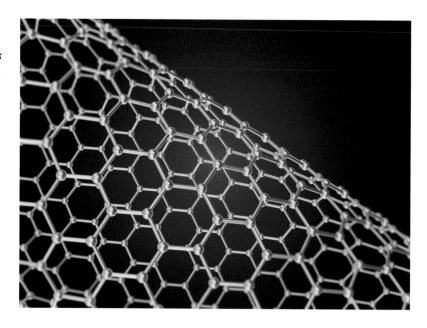

How far can chemistry go in its explanations? We still don't really know. We know that chemistry can explain how a gas behaves, but can it explain how a human brain behaves? The second half of the twentieth century saw the outgrowth of chemistry into biology with tremendous success. We now know details about the structures of the complex molecules at the core of life, and how those structures affect many attributes of living things. We have been able to custom-make molecules to fight disease, and even change the hereditary molecules (DNA) in living organisms to alter the characteristics of those living organisms. The twenty-first century has brought new challenges and new directions. On one frontier, scientists are using the ideas in chemistry to try to explain even more complex phenomena, such as human consciousness, for example. On another frontier, scientists are using chemistry to build ever smaller structures and machines, one atom at a time. Someday we may have molecular submarines, capable of navigating the bloodstream to fight invading cancer cells or viruses. On yet another frontier, scientists have created new materials such as graphene, a two-dimensional substance only one atom thick and stronger than steel. It seems that, at least for the foreseeable future, the power of the particulate model of matter to explain behavior and produce new technology will continue.

A Tour of This Book

In this book, we present the fifty most important ideas in chemistry. Each entry is broken up into several parts: the main entry; the **3-second nucleus** is the idea expressed in a single sentence; the **3-minute valence** describes how the idea fits within a wider context, or can be applied to different circumstances. The book starts with **atoms**, their structures and their properties. It then goes on to show how atoms bond together to form **compounds**, and how we can understand bonding and the resulting **molecules**. From there we move on to the **states of matter** (gases, liquids, and solids) and then on to **chemical reactions**. We then examine **energetics** and describe the laws that govern the flow of energy. Finally, we survey four subfields of chemistry: **inorganic chemistry**, **organic chemistry**, **biochemistry**, and **nuclear chemistry**. Our goal throughout is not to provide exhaustive or detailed accounts of chemistry, but rather to give you a flavor of the field—to show that behind all that happens around you and inside you, particles are doing a complex and beautiful dance that makes it all possible.

The position of electrons within an atom is central to understanding how atoms bond together.

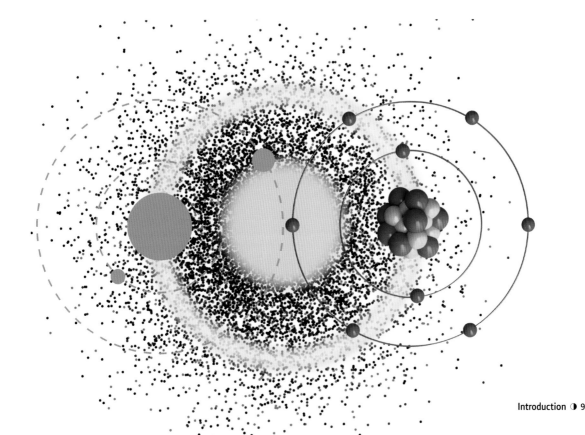

ATOMS, MOLECULES & COMPOUNDS

ATOMS, MOLECULES & COMPOUNDS
GLOSSARY

alkali metals The column of metals (group IA) on the far left of the periodic table that includes lithium, sodium, potassium, rubidium, cesium, and francium.

atomic number The unique number assigned to each element that corresponds to the number of protons in the element's nucleus.

atomic theory The idea that all matter is composed of tiny particles called atoms.

classical physics Physics before the advent of quantum mechanics.

covalent bonding The joining of atoms by the sharing of one or more electrons.

electron A subatomic particle with a negative charge and a mass of 0.00055 amu (atomic mass unit).

element A fundamental substance that cannot be divided into simpler substances. There are ninety-one naturally occurring elements.

Heisenberg's Uncertainty Principle The quantum mechanical principle that certain quantities, such as position and momentum, cannot be simultaneously specified to arbitrary accuracy.

ionic bonding The joining of two atoms by the transfer of an electron from one to the other.

ionic compound A compound, usually composed of a metal and one or more nonmetals, that contains atoms joined by ionic bonds.

isotope An atom that has the same number of protons as another atom, but a different number of neutrons.

mass number The sum of the number of protons and neutrons of an atom.

molecular compound A compound, usually composed of two or more nonmetals, that contains atoms joined by covalent bonds.

neutron A subatomic particle with no charge and a mass of 1 amu.

noble gases The column of gases (group 8A) on the far right of the periodic table that includes helium, neon, argon, krypton, xenon, and radon.

nuclear fusion The joining of two lighter nuclei to form a heavier one.

nuclear model A model for the atom in which most of the mass of the atom is contained in a small dense nucleus composed of protons and neutrons. Most of the volume of the atom is occupied by the electron cloud.

nucleosynthesis The process by which elements form within stars.

proton A subatomic particle with a positive charge and a mass of 1 amu.

quantum mechanics The realm of physics, developed in the early twentieth century, that deals with the very smallest particles that exist.

Schrödinger's Cat thought experiment A thought experiment involving the application of the uncertainty principle to a cat in a box with a radioactive substance that has a 50 percent chance of decaying. If the atom decays, then the cat dies, so the cat is in a strange state of being both dead and undead, with a 50 percent probability of each. Schrödinger used this experiment to show that quantum mechanical ideas are not applicable to large-scale objects (such as cats).

valence electrons The highest energy electrons (and therefore the most important in bonding) in an atom.

velocity A measure of how fast (and in what direction) an object is moving.

MATTER IS MADE OF PARTICLES

3-SECOND NUCLEUS

Matter is composed of particles. The nature of the particles—especially their structure—determines the properties of matter.

3-MINUTE VALENCE

Humans have wondered about the fundamental composition of matter for 2,500 years. The basic question is this: if you divide a sample of matter into smaller and smaller pieces, could you go on forever or would you eventually get to fundamental particles that are no longer divisible? For most of civilization, humans got the answer to this question wrong.

The Ancient Greek philosophers believed that matter was infinitely divisible—that matter had no fundamental particles. Subsequent thinkers followed suit for more than two thousand years. It was not until the eighteenth and nineteenth centuries that early chemists used careful measurements—primarily the relative weights of related samples of matter—to determine otherwise. And it wasn't until the early twentieth century that the question was definitely settled: the 1926 Nobel Prize in Physics was awarded to Jean Perrin for settling the matter. The Greeks were wrong—matter is particulate (it is made up of particles), and those particles are called atoms. And that turns out to be among the most important ideas in all of human thought. Why? Because the idea that matter is made of particles has enabled us to understand nature from the bottom up. What we found was remarkable: as far as we can tell, the particles that compose matter—their composition and structure—determine the properties of matter. Matter does what the particles that compose it do. Water boils at 212°F (100°C) because the three atoms that compose a water molecule bond together in a certain order, at a certain angle, and at certain distances. Change any of these characteristics and water would be a different substance.

RELATED TOPICS

See also
THE STRUCTURE OF THE ATOM
page 16

INSIDE THE ATOM
page 18

WHERE DID ATOMS COME FROM?
page 20

3-SECOND BIOGRAPHIES

JOHN DALTON
1766–1844
English chemist who articulated the atomic theory of matter

JEAN PERRIN
1870–1942
French physicist who studied the motion of tiny particles suspended in liquid to experimentally settle the question of the particulate nature of matter

EXPERT

Nivaldo Tro

Jean Perrin won the Nobel Prize essentially for proving the existence of atoms.

THE STRUCTURE OF THE ATOM

3-SECOND NUCLEUS
An atom consists of a tiny nucleus—containing protons and neutrons—with electrons in a diffuse cloud surrounding the nucleus.

3-MINUTE VALENCE
Matter is particulate—it is made of particles. But what are those particles like? What is their structure? The earliest models implied that the distribution of matter within an atom was fairly uniform, but later experiments suggested otherwise. The atom itself is mostly empty space with nearly all of the mass contained in a small space called the nucleus.

In 1897, J. J. Thomson discovered a new type of particle—the electron—that was much smaller than the atom itself. Thomson demonstrated that electrons were negatively charged, that they were present in all different kinds of matter, and that their mass was one two-thousandth the mass of the lightest atom. Thomson's discovery implied that the atom itself was composed of even smaller particles. Based on his discovery, Thomson developed a model for the atom called the "plum-pudding model." In this model, even the lightest atoms were composed of thousands of electrons held in a sphere of positive charge. In 1909, Ernest Rutherford (pictured) set out to confirm Thomson's model, but he proved it wrong instead. Rutherford accelerated particles (eight thousand times more massive than electrons) at a thin sheet of gold atoms. Most of these particles were not deflected by the gold atoms, but a few bounced back. Rutherford claimed that his results were "about as credible as if you fired a 15-inch [38-cm] shell at a piece of tissue paper and it came back and hit you." Rutherford developed a new model for the atom—the nuclear model—in which the mass of the atom is concentrated in a small space called the nucleus. Most of the volume of the atom is empty space.

RELATED TOPICS
See also
INSIDE THE ATOM
page 18

WHERE DID ATOMS COME FROM?
page 20

THE DUAL NATURE OF THE ELECTRON
page 22

3-SECOND BIOGRAPHIES
J. J. THOMSON
1856–1940
English physicist who discovered the electron

ERNEST RUTHERFORD
1871–1937
New Zealand physicist whose famous Gold Foil Experiment established the nuclear model for the atom

EXPERT
Nivaldo Tro

The nuclear atom, with the nucleus enlarged to be visible. If drawn to scale, the nucleus would be a tiny dot, too small to see.

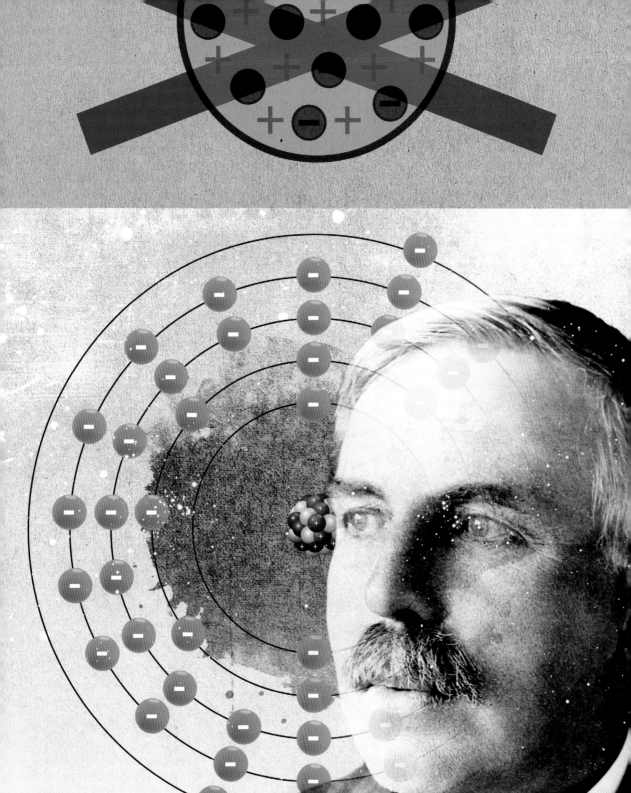

INSIDE THE ATOM

3-SECOND NUCLEUS

An atom is composed of protons, neutrons, and electrons. The number of electrons in a neutral atom always equals the number of protons in its nucleus.

3-MINUTE VALENCE

All atoms are composed of the same three subatomic particles: protons, neutrons, and electrons (see table below). So what makes one atom different from another? The numbers of those particles. Incredible as it may seem, both sodium (a reactive metal that explodes in water) and helium (an inert gas that reacts with nothing) are made of the same subatomic particles, just different numbers of them.

The number of protons in the nucleus of an atom is called the atomic number (Z) and it determines the identity of the atom and its corresponding element. For example, helium (Z=2) has two protons in its nucleus and sodium (Z=11) has eleven protons in its nucleus. The number of known elements ranges from Z=1 to Z=118—as shown in the periodic table on the facing page. Each element has a name, a symbol, and a unique atomic number. The number of neutrons in the nucleus of an atom can vary within atoms of the same element. For example, most helium atoms have two neutrons, but some have three. Atoms with the same number of protons but a different number of neutrons are called isotopes. Since most of the mass of an atom is due to its protons and neutrons, the sum of the numbers of these two particles is called the mass number (A). Scientists specify isotopes with the following notation: $^{A}_{Z}X$, where X is the chemical symbol, Z is the atomic number, and A is the mass number. For example, the helium isotope with 2 neutrons is specified by $^{4}_{2}He$.

Subatomic Particles

	Mass (amu)	Charge (relative)
Proton	1.0	+1
Neutron	1.0	0
Electron	0.00055	-1

RELATED TOPICS

See also
WHERE DID ATOMS COME FROM?
page 20

PERIODIC PATTERNS
page 26

RADIOACTIVITY
page 140

3-SECOND BIOGRAPHY

JAMES CHADWICK
1891–1974
English physicist who discovered the neutron

EXPERT

Nivaldo Tro

The periodic table lists the 118 known elements (91 naturally occurring and 27 synthetic) according to their atomic number (top left in each element box).

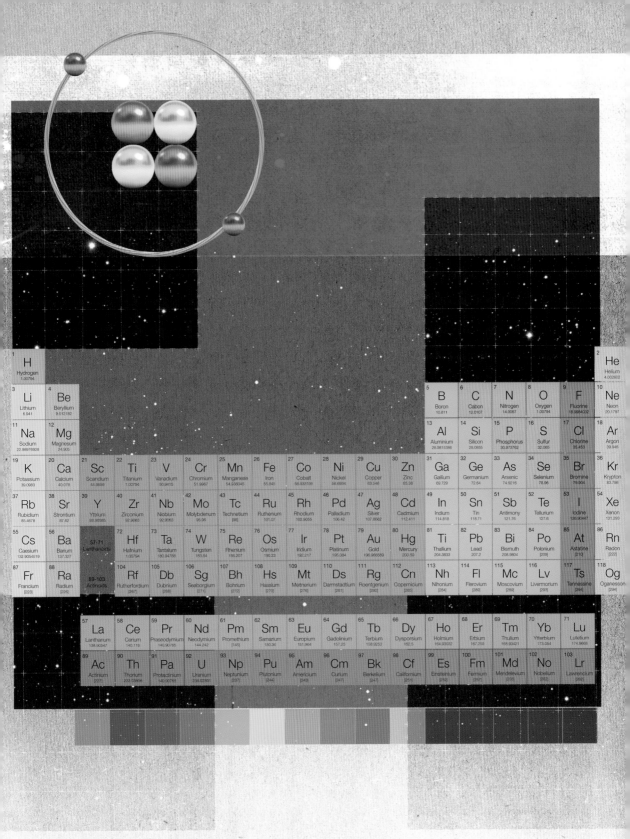

WHERE DID ATOMS COME FROM?

3-SECOND NUCLEUS
Atoms form through
nucleosynthesis, which
began in the first few
minutes after the big bang
and happens to this day
within the core of stars
and supernovae.

3-MINUTE VALENCE
Our planet naturally
contains about 91 different
elements. Where did the
atoms that compose these
elements come from?
How did atoms form? They
formed through a process
called nucleosynthesis,
which began about 13.7
billion years ago at the
very birth of our universe.

According to the big bang model, our universe began as a hot, dense collection of matter and energy that rapidly expanded and cooled. During the first twenty minutes of that expansion, hydrogen and helium (the two most abundant elements in the universe) formed from the soup of subatomic particles. Then nucleosynthesis stopped as the universe continued to expand and cool. Eventually, after about five hundred million years, the first stars formed. Stars are the nurseries in which all other elements are made. As stars burn—through a process called nuclear fusion—they fuse together the nuclei of smaller atoms to form larger atoms. Young stars fuse hydrogen atoms to form helium. This fusion gives off tremendous amounts of heat and light and can power a star for billions of years. As a star ages, and if it is large enough, fusion can continue to form larger atoms such as carbon and oxygen—all the way up to iron. The formation of elements beyond iron requires the input of energy, and only happens in the supernova stage of a star's life. A supernova is essentially a large exploding star. The energy emitted by a supernova can power the nucleosynthesis of elements up to uranium, the heaviest naturally occurring element.

RELATED TOPICS
See also
RADIOACTIVITY
page 140

SPLITTING THE ATOM
page 142

NUCLEAR WEIGHT LOSS
page 144

3-SECOND BIOGRAPHIES
ARTHUR EDDINGTON
1882–1944
English astronomer and
physicist who first suggested
that stars are powered by
fusion

FRED HOYLE
1915–2001
English astronomer who
formulated the theory of
nucleosynthesis within stars

EXPERT
Nivaldo Tro

*In stars, smaller atoms
fuse together to form
larger atoms. All atoms
beyond helium were
born in the core of stars
and supernovae.*

THE DUAL NATURE OF THE ELECTRON

3-SECOND NUCLEUS
For electrons and other small particles, the trajectories of classical physics are replaced with the probability distributions of quantum mechanics.

3-MINUTE VALENCE
Are the smallest particles that exist, such as electrons, just like those that we can see with our eyes, only smaller? Does an electron orbiting an atom behave like a planet orbiting the Sun? No. Electrons behave differently. Electrons, and other small particles, have a wave-particle duality that makes it impossible to predict exact trajectories for them. Instead, we describe their behavior in terms of probability.

An electron traveling through space behaves very differently from a baseball flying toward the outfield. A baseball has a definite trajectory—a deterministic path that it follows. A good outfielder can predict where a baseball will land. This prediction requires that the outfielder simultaneously knows two properties of the flying baseball: its position (where it is) and its velocity (how fast it is going). If the outfielder only knew one of these two properties, he or she could not predict the baseball's path. An electron behaves differently because it has a dual nature: a wave nature (associated with its velocity) and a particle nature (associated with its position). The key to understanding electron behavior is Heisenberg's Uncertainty Principle, which states that "an electron never exists as both a wave and a particle simultaneously." It is either one or the other, but not both. Although Heisenberg's principle solved a great paradox (how something can be both a wave and a particle), it implied the death of determinism. If you can't observe the wave nature and particle nature of the electron simultaneously, then you can't simultaneously know its velocity and its position, which means you can't predict its future path.

RELATED TOPICS
See also
INSIDE THE ATOM
page 18

WHERE ELECTRONS ARE WITHIN AN ATOM
page 24

PERIODIC PATTERNS
page 26

3-SECOND BIOGRAPHIES
ERWIN SCHRÖDINGER
1887–1961
Austrian physicist central to the development of quantum mechanics and known for the thought experiment "Schrödinger's Cat"

WERNER HEISENBERG
1901–76
German physicist who articulated the "Uncertainty Principle"

EXPERT
Nivaldo Tro

In an atom, electrons do not orbit the nucleus like planets orbit the Sun. Instead, they exist in clouds of probability.

WHERE ELECTRONS ARE WITHIN AN ATOM

3-SECOND NUCLEUS

Electrons in atoms exist in quantum mechanical orbitals, three-dimensional probability maps that show the likelihood of finding the electron in a certain volume of space.

3-MINUTE VALENCE

Atoms bond together by sharing or transferring electrons. As a result, the positions of electrons within an atom—where they are—is important because it affects how atoms bond together. In an early model, electrons were thought to orbit the nucleus of the atom much like planets orbit the Sun. However, this model was later proved wrong and was replaced with the quantum mechanical model for the atom.

The electron orbits of early atomic models were later replaced by quantum mechanical orbitals. Unlike a planetary orbit, an *orbital* is a three-dimensional probability map that shows the probability of finding an electron in a certain volume of space. You can understand an orbital with a simple analogy. Imagine taking a photo every ten seconds for several minutes of a moth flying around a light bulb, and then superimposing all the photos to make a single image. The result shows the light bulb with dozens of images of the moth. The volume immediately surrounding the light bulb has many moth images, indicating a high probability of finding the moth in that space. Further away from the light bulb there are fewer moth images, meaning that the probability of finding the moth in that space is lower. A quantum mechanical orbital is analogous—the light bulb is the atomic nucleus and the moth is the electron. Just as the early model of the atom had many different orbits at different distances from the nucleus and with different energies, so the quantum mechanical model has many different orbitals, each with different average distances from the nucleus and with different energies. Electrons can be observed in one orbital or another, but never in between.

RELATED TOPICS

See also
THE DUAL NATURE
OF THE ELECTRON
page 22

BONDING ATOMS TOGETHER
page 28

3-SECOND BIOGRAPHIES

NIELS BOHR
1885–1962
Danish physicist central to the development of the quantum mechanical model for atomic structure

ERWIN SCHRÖDINGER
1887–1961
Austrian physicist central to the development of the quantum mechanical model of the atom

EXPERT

Nivaldo Tro

Early atomic models had electrons orbiting the nucleus like planets orbit the Sun. These have been replaced by the quantum mechanical model.

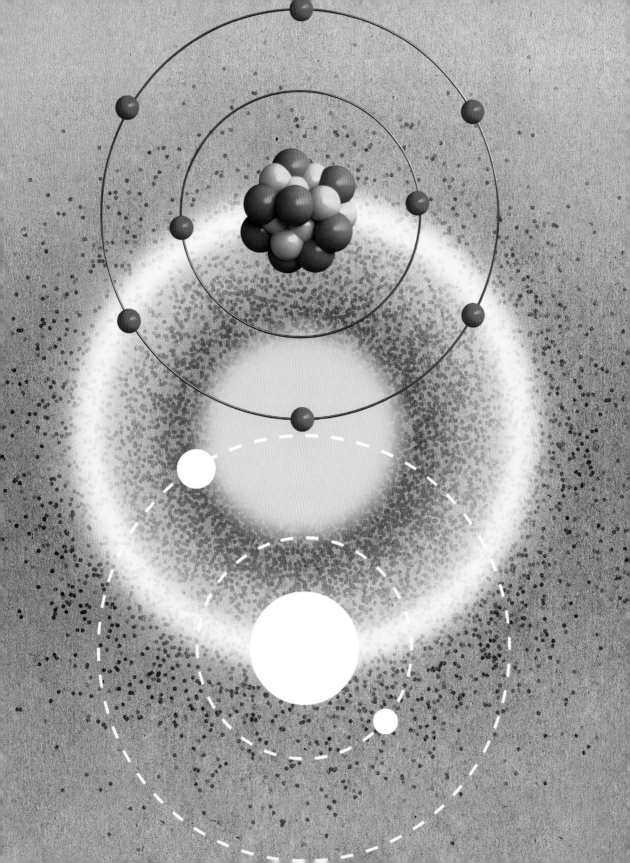

PERIODIC PATTERNS

3-SECOND NUCLEUS
When elements are listed in order of increasing atomic number, their properties recur in a regular pattern.

3-MINUTE VALENCE
Our Earth contains about ninety-one different naturally occurring elements, each one with its own distinctive properties. However, certain groups of elements share similarities. The periodic law and the corresponding periodic table allow us to organize the known elements in ways that help us make sense of their properties.

The ancient Greeks thought that matter was composed of only four elements: earth, water, fire, and air. By the mid-1800s, however, scientists had discovered more than fifty different elements. Dmitri Mendeleev noticed a pattern in the properties of known elements when he listed them in order of increasing atomic number: certain properties recurred periodically. Based on this observation, Mendeleev constructed a table of elements with atomic number increasing from left to right, and elements with similar properties aligning in columns. His table contained some gaps that allowed him to predict the existence and properties of yet undiscovered elements (which were later discovered). Mendeleev's table evolved into the modern periodic table, which lists all known elements to date. The elements on the left and middle of the table are mostly metals, and the elements on the right side of the table are mostly nonmetals. Each column represents a family of elements with similar properties. For example, the far left column contains the alkali metals, a family of elements that are all solid metals at room temperature and highly reactive. The far right column, by contrast, contains the noble gases, a family of elements that are all gases at room temperature and display little or no chemical reactivity.

RELATED TOPICS
See also
INSIDE THE ATOM
page 18

BONDING ATOMS TOGETHER
page 28

THE UNIQUENESS PRINCIPLE
page 86

3-SECOND BIOGRAPHIES
JULIUS LOTHAR MEYER
1830–95
German chemist who made significant contributions to the periodic table

DMITRI MENDELEEV
1834–1907
Russian chemistry professor who formulated the periodic law and constructed one of the first periodic tables

EXPERT
Nivaldo Tro

Mendeleev formulated one of the first periodic tables, which organizes elements according to atomic number and chemical properties.

BONDING
ATOMS TOGETHER

3-SECOND NUCLEUS
Atoms bond together
to form compounds. A
compound, unlike a mixture
of elements, contains two
or more elements in a fixed,
definite proportion.

3-MINUTE VALENCE
The universe contains
118 different elements,
but would be lifeless if
these elements did not
bind together to form
compounds. When two
or more elements combine
to form a compound, a
completely new substance
forms with properties
much different from the
elements that compose it.
In this way, our universe's
118 different elements
can form millions of
compounds. And this,
among other things,
makes life possible.

Atoms bond together by either sharing (covalent bonding) or transferring (ionic bonding) the electrons in their highest-energy orbitals to form compounds. Sharing of electrons typically occurs between two or more nonmetals, resulting in a molecular compound, so called because it is composed of distinct molecules (groups of atoms bonded together). Transfer of electrons typically occurs from a metal to a nonmetal and results in an ionic compound. Ionic compounds do not contain distinct molecules, but rather exist as an array of ions (charged particles) with alternating positive and negative charge. Water is a good example of a molecular compound. We represent a compound with a chemical formula, which tells us the elements present in the compound and the relative number of atoms of each one. For example, the formula for water is H_2O, which means that a water molecule is composed of two hydrogen atoms and one oxygen atom, and the formula for sucrose (table sugar) is $C_{12}H_{22}O_{11}$. Molecular compounds can contain as few as two atoms in a molecule to as many as thousands. Sodium chloride (table salt) is a good example of an ionic compound. The formula for sodium chloride is NaCl, which indicates sodium and chlorine in a one-to-one atomic ratio.

RELATED TOPICS
See also
WHERE ELECTRONS ARE
WITHIN AN ATOM
page 24

THE LEWIS MODEL FOR
CHEMICAL BONDING
page 32

VALENCE BOND &
MOLECULAR ORBITAL
THEORIES
page 34

3-SECOND BIOGRAPHIES
JOSEPH PROUST
1754–1826
French chemist who made
observations on the
composition of compounds

LINUS PAULING
1901–94
American chemist who made
significant contributions to our
understanding of chemical
bonding

EXPERT
Nivaldo Tro

*Water is a molecular
compound, composed
of two hydrogen
atoms bonded to
one oxygen atom.*

1766
Born in Eaglesfield, England, in a white bungalow that still stands

1776
Sent to Pardshaw Hall Quaker School

1781–93
Teaches at the Stramongate School in Kendal, England

1787
Begins to keep meteorological diaries

1793
Now at the New College in Manchester, England, publishes a paper on red-green color blindness, now often known as "daltonism"

1801
Formulates Dalton's Law of Partial Pressures

1803
Delivers a paper in which he first describes his atomic theory

1808
Publishes *A New System of Chemical Philosophy*, in which his atomic theory is presented in full

1810
Nominated for membership of the Royal Society but refuses the offer because Quakers resist public recognition

1822
Renominated and elected to the Royal Society without his knowledge

1826
Receives the Royal Society's first Royal Medal

1832
Initially refuses an honorary doctorate degree from Oxford University that would have required him to wear scarlet robes; persuaded that the robes were a dull green, he accepts the degree

1844
Dies and is accorded a civic funeral with full honors

JOHN DALTON

John Dalton was born into a poor, staunchly Quaker family in the town of Eaglesfield in northern England in 1766. At the age of ten he was sent to a nearby Quaker School and just two years later began teaching there.

Soon afterward he started to teach and study in Kendal, where he began to carry out and record meteorological observations, many with instruments that he built himself. He did this for fifty-seven years until he died, recording more than two hundred thousand observations. He once wrote that "my head is too full of triangles, chymical processes and electrical experiments, etc., to think much of marriage."

Moving to Manchester as a tutor in mathematics and natural philosophy, he joined the Manchester Literary and Philosophical Society. His first communication at the society described a red-green "color blindness"—from which he and his brother suffered—that is still referred to as "daltonism."

His love of meteorology led him to consider the composition of the air and the nature of its component gases. He concluded that the atmosphere is a mixture of gases and that the total pressure it exerts is the sum of the "partial pressures" that each individual gas exerts. The overall pressure, he maintained, is due to the particles (what we would call atoms and molecules) of these gases slamming against the walls of the container in which they are held.

Most famously, Dalton devised the first concrete atomic theory that organized a number of the assumptions known at the time, starting with the Greek idea of indivisible atoms. He maintained that the atoms of a given element are unique (particularly in mass) and combine with atoms of another element in whole-number ratios. In chemical reactions, atoms are reshuffled from one configuration to another. He went on to construct one of the first ever tables of atomic weights but this was marred by inaccuracies that could have been easily corrected had he been more open to valuable new ideas from the international scientific community. Despite these difficulties his theory could be tested and its general assumptions held up well.

Given that Dalton had taken a concept that had been imprecisely discussed and largely rejected for two thousand years and fashioned it into a guiding paradigm that revolutionized all of science, his colleagues were anxious to honor him. However, due to his Quaker beliefs he refused many of these honors, including a doctorate from Oxford University that would have required him to wear scarlet robes. Later, told that the robes were in fact green (he was color-blind, after all), he received the degree. When Dalton died, he could not stop forty thousand people filing past his coffin and one hundred coaches following his funeral cortège.

Glen E. Rodgers

THE LEWIS MODEL FOR CHEMICAL BONDING

3-SECOND NUCLEUS
In the Lewis model for chemical bonding, atoms bond to obtain octets— eight electrons in their valence shell.

3-MINUTE VALENCE
The most powerful pieces of scientific knowledge are theories (or models). Theories explain not only what happens in nature but also why it happens. The Lewis model for chemical bonding explains why, for example, water is H_2O and not some other combination of atoms. The Lewis model is simple, however, and other more sophisticated models are even more powerful at predicting and explaining chemical bonding.

In the simplest model for chemical bonding, called the Lewis model, atoms share or transfer their highest energy electrons (called valence electrons) to obtain an octet—eight electrons in their highest energy (or outermost) set of orbitals. One important exception is hydrogen, which shares/transfers its one electron to obtain a duet—two electrons in its outermost orbital. When applying the Lewis model, chemists use special symbols to represent atoms and their valence electrons. For example, the Lewis symbols for hydrogen and oxygen are as follows:

$$H\cdot \quad \cdot \ddot{O}\cdot$$

The one dot next to H represents hydrogen's one valence electron and the six dots around the O represent oxygen's six valence electrons. The bonding between hydrogen and oxygen to form water involves the sharing of the valence electrons, and we draw the Lewis symbol for water as follows:

$$H:\ddot{O}:H \qquad \text{Duet}\underbrace{\overbrace{(H{:}O{:}H)}}_{\text{Octet}}\text{Duet}$$

The shared electrons (those between two elements) count toward the octet (or duet) of both atoms. So in this Lewis structure, each hydrogen atom has a duet and oxygen an octet.

RELATED TOPICS
See also
WHERE ELECTRONS
ARE WITHIN AN ATOM
page 24

BONDING ATOMS TOGETHER
page 28

VALENCE BOND &
MOLECULAR ORBITAL
THEORIES
page 34

3-SECOND BIOGRAPHY
GILBERT N. LEWIS
1875–1946
American chemist and University of California, Berkeley, chemistry professor who constructed the Lewis model for chemical bonding

EXPERT
Nivaldo Tro

The Lewis model shows how atoms share electrons to obtain octets.

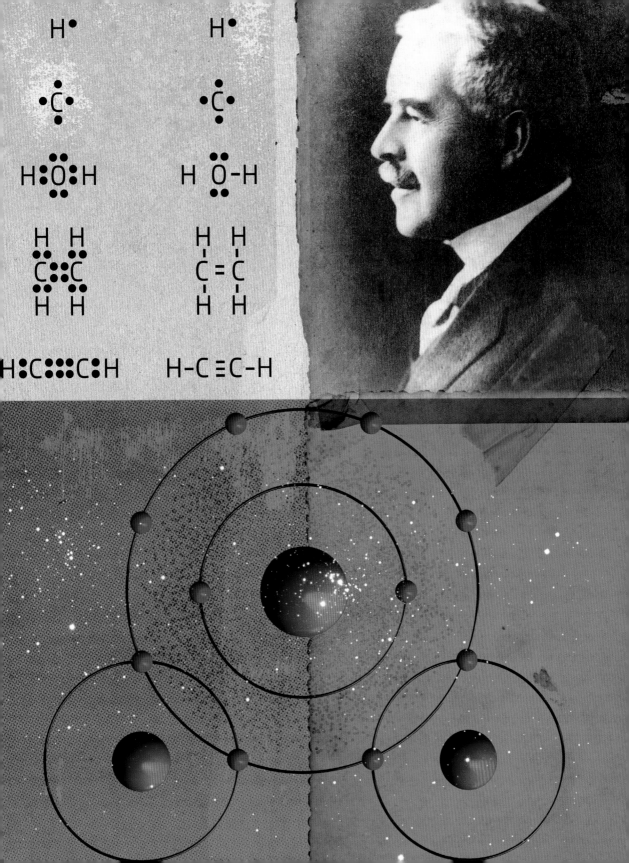

VALENCE BOND & MOLECULAR ORBITAL THEORIES

3-SECOND NUCLEUS

In the valence bond model, a chemical bond is the overlap between half-filled atomic orbitals. In molecular orbital theory, atomic orbitals are completely replaced by molecular orbitals.

3-MINUTE VALENCE

The Lewis model for chemical bonding is practical and useful; however, it also has limits. We know, for example, that electrons are not stationary dots that sit between atoms. Two more powerful bonding models—valence bond theory and molecular orbital theory—take into account the quantum mechanical nature of electrons and provide even more powerful predictions and explanations of chemical bonding.

Chemists use three different models to explain chemical bonding: the Lewis model, valence bond theory, and molecular orbital theory, each increasingly complex but also increasingly powerful. The Lewis model requires nothing more than paper and pencil to enable chemists to predict and explain a great deal of chemical behavior. Valence bond and molecular orbital theory, by contrast, both require more complex calculations, usually on a computer. In valence bond theory, a chemical bond is modeled as the overlap between half-filled atomic orbitals. As the orbitals overlap, the energy of the electrons in those orbitals decreases, stabilizing the molecule relative to its constituent atoms. A molecular orbital is to a molecule what an atomic orbital is to an atom. Each molecule has its own unique set of molecular orbitals that depend on the constituent atoms and their arrangement in space. If the overall energy of the electrons in the molecular orbitals is lower than in the constituent atoms' atomic orbitals, the resulting molecule is stable. Both valence bond theory and molecular orbital theory can accurately predict details about molecular structure including molecular geometries, bond lengths, and bond strengths.

RELATED TOPICS

See also
WHERE ELECTRONS ARE WITHIN AN ATOM
page 24

BONDING ATOMS TOGETHER
page 28

THE LEWIS MODEL FOR CHEMICAL BONDING
page 32

3-SECOND BIOGRAPHIES

JOHN EDWARD JONES
1894–1954
English mathematician, physicist, and pioneer of computational chemistry

LINUS PAULING
1901–94
American chemist who made significant contributions to valence bond theory

EXPERT

Nivaldo Tro

Molecular orbital theory predicts that oxygen should be a magnetic liquid, which it is. The simpler bonding theories fail to predict this property.

OPPOSITES ATTRACT

3-SECOND NUCLEUS

The often uneven distribution of electrons that can result when two different atoms bond together results in a polar bond, which greatly affects a substance's properties.

3-MINUTE VALENCE

The existence of liquid water on Earth's surface can be attributed to polar bonds. Most small molecules are gases at room temperature, but water is among the very few that is a liquid. Why? Because water has highly polar bonds with hydrogen at one end and oxygen on the other. The small size of hydrogen allows the molecules to get very close together and interact strongly. This strong interaction makes it difficult to separate water molecules from one another.

We know from previous entries that atoms can bond together by sharing electrons. But if the bonding atoms are different (two different elements), then the sharing is often not equal—one of the two atoms hogs the electron more than the other. The result is a polar bond, one that has a positive end on one side and a negative end on the other. In a molecule, polar bonds may add together to result in a polar molecule. Polar molecules interact strongly with one another because the positive end of one molecule is attracted to the negative end of its neighbor— just as the north pole of a magnet is attracted to the south pole of another magnet. These attractions affect the properties of the substances that the molecules compose. For example, polar substances tend to have higher melting and boiling points than their nonpolar counterparts because the attraction between neighboring molecules makes the molecules more difficult to separate. Polar substances also tend not to mix well with nonpolar substances. For example, water and oil do not mix because water is very polar and oil is nonpolar. Water and ethyl alcohol (ethanol), by contrast, mix in all proportions because they are both polar.

RELATED TOPICS
See also
BONDING ATOMS TOGETHER
page 28

THE FORCES THAT HOLD
MATTER TOGETHER
page 42

THE LIQUID STATE
page 46

3-SECOND BIOGRAPHIES
JOHANNES DIDERIK
VAN DER WAALS
1837–1923
Dutch physicist who was among the first to postulate forces between molecules

LINUS PAULING
1901–94
American chemist who quantified the polarity of chemical bonds

EXPERT
Nivaldo Tro

A polar molecule has an asymmetrical charge distribution that causes an attraction to other polar molecules.

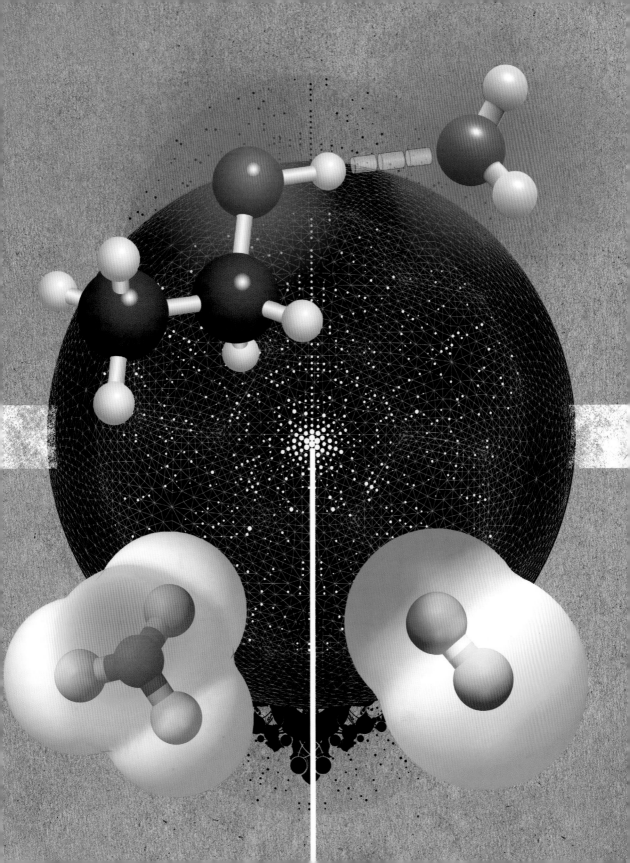

STATES OF MATTER

amorphous solid A solid whose atoms or molecules are not arranged in an ordered, repeating, three-dimensional array.

covalent bond The joining of atoms by the sharing of one or more electrons.

crystalline solid A solid whose atoms or molecules are arranged in a well-ordered, repeating, three-dimensional array.

dipole force The attractive force that exists between two or more polar molecules due to an asymmetric distribution of charge.

dispersion force The attractive force that exists between atoms and molecules due to temporary dipoles that develop because of charge fluctuations.

homogenous mixture A mixture containing two or more components that has the same composition throughout.

intermolecular forces Attractive forces that exist between atoms and molecules.

ionic bond The joining of two atoms by the transfer of an electron from one to the other.

"like dissolves like" principle The idea that polar molecules mix well with other polar molecules, but don't mix well with nonpolar molecules. Mostly applicable to aqueous solutions.

macroscopic properties of gases Properties such as temperature, volume, pressure, and number of moles of a sample of a gas.

osmotic pressure The pressure required to stop osmotic flow. Osmotic flow is the flow of water through a semipermeable membrane from a more dilute solution to a more concentrated one.

polar bond A chemical bond with asymmetric charge distribution.

polar molecule A molecule containing polar bonds that also has an asymmetric charge distribution over the entire molecule.

polar solute A solute (component of a solution) that has an asymmetric charge distribution. A nonpolar solute has a highly symmetrical charge distribution.

single liquid phase A liquid mixture with uniform composition.

solute The minority component of a solution.

solvent The majority component of a solution.

vapor pressure The pressure of the vapor of a liquid in equilibrium with its liquid.

THE FORCES THAT HOLD MATTER TOGETHER

3-SECOND NUCLEUS

Solid and liquid matter exist because the particles that compose them have strong attractions to one another.

3-MINUTE VALENCE

Why are some substances solid at room temperature, while others are liquids or gases? Because the particles that compose matter are attracted to one another in varying degrees. Strong attractions between particles are responsible for solids at room temperature, moderately strong attractions result in liquids, and weak attractions result in gases. The higher the temperature, the stronger the attractions between particles must be to maintain the liquid and solid state.

Matter exists in three states: solid, liquid, and gas. In the gas state, the particles that compose matter are separated by large distances and do not interact with one another very much. In the solid and liquid states, by contrast, the particles interact strongly, held together by attractive forces. Some solids, such as a diamond, are held together by covalent chemical bonds between atoms (which is what makes diamond so strong). Other solids, such as table salt, are held together by ionic bonds between ions. Still other solids, such as ice, and many liquids are held together by attractive forces that exist between molecules. These forces are known as intermolecular forces. Intermolecular forces exist because the electron distribution in a molecule can be either temporarily asymmetrical, resulting in the dispersion force, or permanently asymmetrical, resulting in the dipole force. In either case, the asymmetrical electron distribution causes part of the molecule to be positively charged (either temporarily or permanently) and another part to be negatively charged. The positive and negative ends of neighboring molecules are then attracted to one another much like opposite poles of a magnet are attracted to one another. These attractions must be overcome for a substance to melt or boil.

RELATED TOPICS

See also
OPPOSITES ATTRACT
page 36

THE LIQUID STATE
page 46

THE SOLID STATE
page 48

3-SECOND BIOGRAPHY

JOHANNES DIDERIK
VAN DER WAALS
1837–1923
Dutch physicist who was among the first to postulate forces between molecules

EXPERT

Nivaldo Tro

The strength of the intermolecular forces between water molecules in ice determines its melting point.

THE GASEOUS STATE

Gases have the unique property

—unlike solids and liquids—of always completely filling the volume of their container. By the end of the eighteenth century, the relationships between the volume, pressure, and amount of gas had been empirically described. Hot-air balloonists Jacques Charles (of Charles's law fame) and Joseph Gay-Lussac not only set new altitude records for ballooning, but also used these adventures to collect data on the temperature-volume relationship of gases, which indicated that the volume occupied by a gas increased with temperature. Robert Boyle demonstrated that the volume occupied by a gas is inversely proportional to its pressure, a relationship that became known as Boyle's law. Amedeo Avogadro hypothesized that equal volumes of gas were occupied by equal numbers of molecules as part of his theory that gases consisted of molecules that in turn were composed of atoms. His theory was largely ignored at the time. Further progress in understanding the origin of these relationships required the acceptance by chemists of the theory of the particle nature of atoms and molecules. The current model to explain these properties is kinetic-molecular theory.

3-SECOND NUCLEUS
Kinetic-molecular theory explains the macroscopic properties of gases based on the behavior of gas particles.

3-MINUTE VALENCE
Kinetic-molecular theory is based on three postulates. The sizes of the particles that comprise a gas are negligibly small, so that gas particles occupy essentially none of the volume of a gas. Gas particles are in constant motion and have an average kinetic energy proportional to the temperature of the gas. Collisions between gas molecules are perfectly elastic in that energy can be transferred but not lost in the collisions.

RELATED TOPICS
See also
MATTER IS MADE OF PARTICLES
page 14

ENTROPY & THE SECOND LAW OF THERMODYNAMICS
page 76

3-SECOND BIOGRAPHY
JAMES CLERK MAXWELL
1831–79
Scottish mathematician who, while best known for his work in electromagnetism, also worked on developing a statistical means of explaining the properties of gases

EXPERT
John B. Vincent

The properties of gases make both hot-air ballooning and scuba diving possible.

THE LIQUID STATE

3-SECOND NUCLEUS
Liquids are made up of molecules with enough energy to flow past one another, but generally not enough to overcome their mutual attraction entirely.

3-MINUTE VALENCE
Pour water into a glass—it fills to a certain level, and the shape of the water is the same as the inside of the glass. If you pour the water into a square-shaped glass, it assumes that shape. If you leave the water to stand for a few days, it slowly evaporates away. How do we explain this behavior from a particulate viewpoint?

The molecules that compose a liquid are like dancers in a crowded night club. The dancers have so much energy that they move around the floor constantly interacting with different people. They are attracted to everyone else in the club and want to dance with everyone. Similarly, molecules in liquids have attractive forces with all the other molecules around them, but they have so much energy that they don't stay still and are constantly moving past one another. As a whole, the people in the dance club have the same shape as the club. If they all moved from a square club to a circular one, their overall shape would change. Similarly, as a whole, water molecules flow to assume the shape of their container. Not every molecule in a liquid (or person in a dance club) has the same amount of energy. Some have more, some have less. A few have a lot more energy—so much that they can break free of their attraction to the other molecules in the liquid and fly out on their own as a gas molecule (they essentially dance right out of the club). This is how liquids evaporate.

RELATED TOPICS
See also
MATTER IS MADE
OF PARTICLES
page 14

THE FORCES THAT HOLD
MATTER TOGETHER
page 42

THE GASEOUS STATE
page 44

3-SECOND BIOGRAPHIES
ROGER JOSEPH BOSCOVICH
1711–87
Ragusan (now Croatia) physicist who predicted that the states of matter depended on forces between their particles

FRANÇOIS-MARIE RAOULT
1830–1901
French chemist who explored the properties of solutions

EXPERT
Jeff C. Bryan

The molecules that compose the liquid state are in constant motion, not unlike dancers in a crowded night club.

THE SOLID STATE

The particles that compose a

solid are like the dancers in the nightclub from the previous entry (on liquids), except that they have less energy relative to the strength of their attractions. The dancers are still shaking, but they are not moving around each other because they are strongly attracted to those currently surrounding them. Similarly, the attractions between the particles in a solid are so strong compared to the energy they possess that the particles don't flow past one another as they do in a liquid. The particles that compose a crystalline solid are not only stuck in one place, but also arranged in an orderly fashion like bricks in a wall. In contrast, the particles that compose an amorphous solid are arranged in a more haphazard way, like a pile of macaroni. Crystalline solids (such as salt or ice) tend to be less flexible than amorphous solids (such as plastic or glass). Although the particles that compose solids do not move past or around one another, they do wiggle and shake. When heated, they get more energy. Eventually, when heated enough, they start moving past one another and the solid melts. The amount of energy (temperature) needed for melting depends on how strongly the molecules are attracted to each other.

3-SECOND NUCLEUS
Solids have a definite size and shape because the particles that compose them are stuck in place.

3-MINUTE VALENCE
Solids behave differently from liquids or gases. They have a fixed shape and size and don't assume the shape of their container like a liquid, nor are they compressible like a gas.

RELATED TOPICS
See also
MATTER IS MADE
OF PARTICLES
page 14

THE FORCES THAT HOLD
MATTER TOGETHER
page 42

THE LIQUID STATE
page 46

3-SECOND BIOGRAPHIES
WILLIAM LAWRENCE BRAGG
1890–1971
Australian-born British physicist and winner of the 1915 Nobel Prize in Physics who discovered how to peer in at the structures of solids

LINUS PAULING
1901–94
American winner of the 1954 Nobel Prize in Chemistry, who developed our understanding of how atoms and molecules are attracted to each other

EXPERT
Jeff C. Bryan

The molecules in the solid state are like dancers who are stuck in one place on the dance floor.

1627
Born in Lismore, Ireland

1639
Sent on a Grand Tour
to Paris, Geneva, and
Florence

1644
Returns to England after
his father's death and
stays with his sister
Katherine, Lady Ranelagh
in London, where they
are members of the
"Invisible College"

1645
Moves into his father's
manor in Stalbridge in
Dorsetshire, England,
where he sets up his first
laboratory

1654
Moves to Oxford,
where he again lives
with his sister; he and his
assistant Robert Hooke
establish a laboratory.
Hooke builds a
"pneumatic engine"

1661
Publishes *The Sceptical
Chymist*, in which he
strongly expresses his
corpuscular or atomic
hypothesis and provides a
definition of an element

1662
Publishes the second
edition of *The Spring
of the Air*, in which he
establishes Boyle's Law

1668
Moves back to London,
where he and Hooke
again establish a
laboratory; by now the
"Invisible College" has
become the "Royal
Society of London for
Improving Natural
Knowledge"

1691
Dies in London one week
after his sister

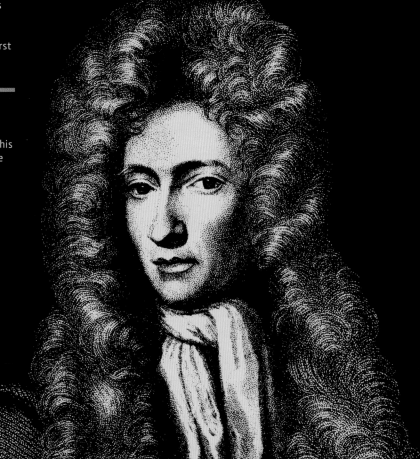

ROBERT BOYLE

Robert Boyle, born in a castle in Lismore, Ireland, in 1627, was the fourteenth child of Richard Boyle, the Great Earl of Cork and one of the richest men in Britain. Boyle's father sent his 12-year-old son on a Grand Tour of Paris, Geneva, and Florence that kept him safe during the Irish Rebellion of 1641. In Geneva, during a fearsome, life-threatening thunderstorm, he pledged to devote his life to promoting God's work on Earth. His ensuing religiosity continued for the rest of his life.

He returned to London to live with his sister Katherine (the "Lady Ranelagh") in 1644, and together they became early members of the "Invisible College," so named because it did not have a regular meeting place, which met to discuss science—then still known as "natural philosophy." In Stalbridge, Dorsetshire, convinced that observations and experiments were the cornerstones on which scientific investigations had to be built, he set up his first laboratory.

Boyle is properly regarded as a transitional figure between alchemy and chemistry. He was responsible for dropping the "al-" of alchemy and is best referred to as a "chymist." Among many other interests, he was devoted to chrysopoeia, the science of transmutating "base metals" into gold. Starting in 1654, he and his assistant Robert Hooke set up a laboratory in Lady Ranelagh's house in Oxford. Hooke built a "pneumatic engine" or vacuum pump that Boyle used to establish what is today known as Boyle's Law, which states the inverse relationship between the volume of a gas and the pressure exerted on it. His experiments lent great support to the idea that air was composed of discrete, rapidly moving "corpuscles" (what we would call atoms or molecules) that collide with the walls of their container and exert a pressure that Boyle called "the spring of the air."

In his book *The Sceptical Chymist* (1661) he defined elements as "certain primitive and simple, or perfectly unmingled bodies." He argued that the Aristotelian view that the elements should be confined to earth, air, fire, and water was not supported by observations. His primary goal was to transform alchemy to make it more scientific by framing good hypotheses based on sound experimental methods reported with great accuracy and detail. In London, Boyle established another laboratory and, among other endeavors, was active in the isolation and production of the new element phosphorus. He was also a prolific writer of more than forty books on a variety of topics, including chemistry as well as philosophy, medicine, and religion. Boyle died in 1691, one week after his beloved sister.

Glen E. Rodgers

CERAMICS

RELATED TOPICS
See also
THE LEWIS MODEL FOR
CHEMICAL BONDING
page 32

THE SOLID STATE
page 48

3-SECOND NUCLEUS
The technologically
useful properties of
ceramic materials depend
on the 3D arrangement of
their atoms and the nature
of the chemical bonds
holding them together.

3-MINUTE VALENCE
The first ceramic figurines
and pots were made more
than twenty thousand
years ago, long before
metal tools. Later artisans
used tough ceramics like
porcelain, clear ceramics
like glass, and the cements
that dominate cityscapes
today. Ceramic scientists
continue to produce new
technologically useful
materials: recent examples
include silicon carbide
cutting tools, boron-nitride
lubricants, silicon computer
chips, and bioglass-based
medical implants made
from silica and
hydroxyapatite.

Ceramics are among the most important materials in human civilization. They are solids held together by networks of ionic or covalent bonds extending throughout the material. They differ from metals in that the bonds are to some extent directional and so must be broken in order for planes of atoms to slip past one another. As a result, ceramics cannot be easily deformed, as metals can, but instead tend to be brittle and hard. Ceramics may be made by mixing finely powdered minerals and heating them until their atoms are moving fast enough to move into each other or until the minerals melt into a single liquid phase. When cooled, the atoms in the resulting ceramic are often aligned in the neat rows of a crystal. Glasses may be produced if the cooling is fast enough so that the atoms are frozen in a snapshot of the chaotic liquid from which they were made. Many aluminosilicate mineral ceramics consist of covalently bonded chains or sheets that can absorb water and metal ions in between the layers. These include clays that swell considerably when they take up water, as well as the mineral kaolinite, which is a primary component of fine china or porcelain.

3-SECOND BIOGRAPHIES
HERMANN SEGER
1839–93
German chemist who pioneered
the scientific study of ceramics
using the periodic table

RUSTUM ROY
1924–2010
Indian-born scientist who
developed the sol-gel method
for preparing ceramics from
liquid chemical precursors

W. DAVID KINGERY
1926–2000
American material scientist
who first applied solid-state
chemistry principles to ceramic
synthesis and processing

EXPERT
Stephen Contakes

*The structure and bonding
in ceramic materials
determine their many
useful properties.*

SOLUTIONS

Solutions are homogenous

mixtures formed when one substance (a solute) dissolves in another substance (a solvent). Ocean water, air, and sugar water are common examples of solutions. Aqueous solutions are those in which water is the solvent. Solutions show different properties from the components that compose them. For instance, a salt-water solution has a lower freezing temperature than pure water (which is one reason why freshwater lakes freeze more easily than oceans). Similarly, a salt-water solution has a higher boiling temperature, lower vapor pressure, and higher osmotic pressure when compared to pure water. Such solution characteristics are known as "colligative properties." They were first studied experimentally by Richard Watson, a Professor of Chemistry at Cambridge University, who observed the freezing time of a series of eighteen aqueous solutions of various salts exposed to an unusually cold (6.8°F/-14°C) February atmosphere in Cambridge. He realized that the primary factor determining the degree of lowering in the freezing point of a solution is the number of solute particles (concentration) and not the type of salt. Calcium chloride ($CaCl_2$) is more effective in treating icy roads than sodium chloride (NaCl) because it provides more solute particles (ions) when mixed with the icy surface.

3-SECOND NUCLEUS
When the particles of one substance are dissolved in a second (the solvent), they interfere with the way the solvent molecules interact, changing the properties of the solution.

3-MINUTE VALENCE
The nature of intermolecular forces in all states of matter partly determines whether one substance dissolves in another. The "like dissolves like" principle is helpful in determining solubility in water: polar solutes tend to be most soluble in water (since water is polar). For example, salt is soluble in water but grease (mostly nonpolar) is not.

RELATED TOPICS
See also
THE GASEOUS STATE
page 44

THE LIQUID STATE
page 46

3-SECOND BIOGRAPHIES
DAVID BERNOULLI
1700–82
Swiss mathematician whose work *Hydrodynamica* provided the first qualitative discussion of aqueous salt solutions

RICHARD WATSON
1737–1816
Professor of Chemistry at the University of Cambridge who first carried out experimental measurements studying the properties of aqueous salt solutions

EXPERT
Ali O. Sezer

Salt water solutions have a lower freezing point than pure water, which is why salt is used to reduce snow and ice build-up on roads and pathways.

CHEMICAL REACTIONS & ENERGETICS

acid A substance that produces H^+ ions in solution. Acids neutralize bases, producing water as a product.

base A substance that produces OH^- ions in solution. Bases neutralize acids, producing water as a product.

catalyst A substance that increases the rate of a reaction without being consumed by the reaction.

chemical energy The energy that can be obtained from a chemical reaction when the reactants have greater potential energy than the products.

chemical reaction A process in which the atoms in one or more substances (the reactants) rearrange to form different substances (the products).

electrolysis The use of electrical current to drive a chemical reaction that would not happen spontaneously.

electrolyte A substance that produces an electrically conductive solution when dissolved in water.

entropy A thermodynamic quantity related to the amount of energy dispersed into a substance at a given temperature.

enzyme A protein that acts as a biological catalyst to increase the rate of a biochemical reaction.

exothermic A reaction that emits energy into the surroundings.

filtration A process of separation in which a solid is separated from a liquid using a filtration device such as a funnel and filter paper.

greenhouse gas An atmospheric gas that is transparent to visible light, but absorbs infrared light. These atmosphere gases act like glass in a greenhouse, allowing light to enter but preventing heat energy from escaping. The three most important greenhouse gases in Earth's atmosphere are water vapor, carbon dioxide, and methane.

hydrocarbon An organic compound containing only carbon and hydrogen.

Kelvin scale An absolute scale used for measuring temperature. On the Kelvin scale, water freezes at 273 K and boils at 373 K. The lowest possible temperature (at which point molecular motion stops) is zero on the Kelvin scale.

kinetic energy The energy associated with the motion of an object or particle.

neutralization A chemical reaction between an acid and a base that typically produces water and a salt.

oxide An oxygen-containing compound.

potential energy The energy associated with the position (within a field) or composition of an object.

precipitation A reaction between two solutions in which a solid (or precipitate) forms.

reactant Any one of the substances that undergoes a chemical reaction. In a reaction, reactants react to form products.

thermal energy The energy associated with the random thermal motion of atoms and molecules.

thermodynamics The study of energy and its conversions from one form to another.

transition-metal oxides Compounds containing a transition metal and oxygen.

voltaic cell A chemical cell that employs a spontaneous chemical reaction to produce electrical current.

CHEMICAL EQUATIONS

3-SECOND NUCLEUS

A chemical equation is a way to precisely represent a chemical reaction, a change in which the atoms that compose one or more substances rearrange to form one or more different substances.

3-MINUTE VALENCE

Chemical reactions occur all around us all the time. For example, our cars are powered by chemical reactions; cooking is a chemical reaction; and our bodies maintain a myriad of reactions that allow us to think, move, eat, and reproduce. Chemical equations not only represent the identities of the reactants and products in a chemical reaction; they also give us quantitative relationships between the amounts that react.

Chemists must move seamlessly

between three related worlds: the macroscopic world that exists in the lab within beakers, flasks, and test tubes; the atomic and molecular world, which we can't see but are constantly trying to imagine and understand; and the symbolic world, which is how we represent the atomic and molecular world on paper. A chemical equation is a way to symbolically represent changes that occur in the atomic and molecular world. These changes are called chemical reactions and they often (although not always) result in significant changes in the macroscopic world. For example, the burning of natural gas is a chemical reaction. In this reaction methane gas combines with oxygen to form carbon dioxide and water. In the macroscopic world, we see the reaction as a blue flame on our stove top. In the molecular world, methane molecules combine with oxygen molecules and transform into carbon dioxide molecules and water molecules. In the symbolic world, we represent the reaction with the following chemical equation:

$$CH_4 + 2\,O_2 \rightarrow CO_2 + 2\,H_2O.$$

Chemical equations must be balanced: they must contain the same number of each type of atom on either side of the equation. Why? Because in a chemical reaction, matter is conserved. Atoms can't just vanish or form out of nothing.

RELATED TOPICS

See also
BONDING ATOMS TOGETHER
page 28

COMBUSTION REACTIONS
& ENERGY SOURCES
page 62

NEUTRALIZING: ACIDS
& BASES
page 64

3-SECOND BIOGRAPHIES

ROBERT BOYLE
1627–91
Anglo-Irish chemist who formulated some of the earliest ideas about chemical reactions

ANTOINE LAVOISIER
1743–94
French chemist who contributed significantly to our understanding of chemical reactions, especially combustion

EXPERT

Nivaldo Tro

We witness chemical reactions, such as the burning of natural gas, every day.

COMBUSTION REACTIONS & ENERGY SOURCES

The first chemical reaction that early humans used was burning or combustion. In a combustion reaction, a substance combines with oxygen and most commonly produces carbon dioxide, water, and other oxide products. Combustion reactions are useful because they emit heat as they occur—they are exothermic. Certain molecules, especially hydrocarbons, have an inherently high potential energy that can be released through combustion. For this reason, our society's fuels are largely based on hydrocarbons. Natural gas is primarily methane (CH_4). Liquefied petroleum (LP) gas is a mixture of propane (C_3H_8) and butane (C_4H_{10}). Petrol is a mixture of hydrocarbons, such as octane (C_8H_{18}), containing five or more carbon atoms. Coal is also a major part of our energy equation: it is mostly carbon and combines with oxygen to form carbon dioxide. Together these fuels are known as fossil fuels because they originated from ancient plant and animal life. However, the combustion of fossil fuels is not without problems. The most vexing problem is probably the emission of carbon dioxide, a greenhouse gas that is affecting Earth's climate. Since the Industrial Revolution, atmospheric carbon dioxide levels have risen by about 38 percent, and average global temperatures have risen by about 1.4°F (0.8°C).

RELATED TOPICS
See also
CHEMICAL EQUATIONS
page 60

HYDROCARBONS
page 106

3-SECOND BIOGRAPHY
ANTOINE LAVOISIER
1743–94
French chemist who contributed significantly to our understanding of chemical reactions, especially combustion

EXPERT
Nivaldo Tro

Combustion or burning is common in energy generation and industry.

NEUTRALIZING: ACIDS & BASES

Most of us experience acids by

taste—they are sour. Citrus fruits, vinegar, carbonated beverages, yogurt, and sour sweets all owe their tangy deliciousness to the acids they contain. Chemists prefer not to taste their work and often define acids as chemicals that produce the hydrogen ion (H^+) when dissolved in water. If acids are a chemical yin, then bases are the yang. Bases tend to taste bitter and produce hydroxide (OH^-) in water, the chemical opposite of H^+. When an acid and base are mixed, the hydrogen ion combines with the hydroxide ion to form HOH (water):

$$H^+ + OH^- \rightarrow H_2O.$$

This type of chemical reaction is called neutralization. If exactly equal amounts of H^+ and OH^- are mixed, the resulting solution will contain neither (because all of the H^+ and OH^- ions reacted to form water), and it won't be acidic or basic. Our stomachs use hydrochloric acid (HCl) to help digest our food. If we eat too much, especially acidic or fatty foods, our stomachs can produce too much HCl, causing us to feel uncomfortable (sometimes called "sour stomach"). To neutralize the excess acid, we can take an antacid. Antacids are bases like aluminium hydroxide ($Al(OH)_3$), magnesium hydroxide ($Mg(OH)_2$), and calcium carbonate ($CaCO_3$) that neutralize the excess stomach acid.

3-SECOND NUCLEUS
Acids produced H^+ in water while bases produce OH^-. An acid and a base react to produce water, effectively neutralizing each other.

3-MINUTE VALENCE
The pH scale is used to measure the acidity or basicity of a solution. The lower the pH value, the higher the concentration of H^+, and the more acidic the solution. The higher the pH value, the higher the concentration of OH^-, and the more basic the solution. Pure water is neutral and has a pH of 7. Stomach acid is pH 1.6, tomato juice 4.2, seawater 8.2, and milk of magnesia ($Mg(OH)_2$) 10.4.

RELATED TOPICS
See also
OPPOSITES ATTRACT
page 36

SOLUTIONS
page 54

CHEMICAL EQUATIONS
page 60

3-SECOND BIOGRAPHIES
SVANTE ARRHENIUS
1859–1927
Swedish winner of the Nobel Prize for Chemistry (1903), who first suggested the acid/base definitions given here

JOHANNES BRØNSTED
& MARTIN LOWRY
1879–1947 & 1874–1936
Danish and British chemists who defined acids and bases on how they react

EXPERT
Jeff C. Bryan

Antacids contain bases that neutralize stomach acid, the cause of heartburn.

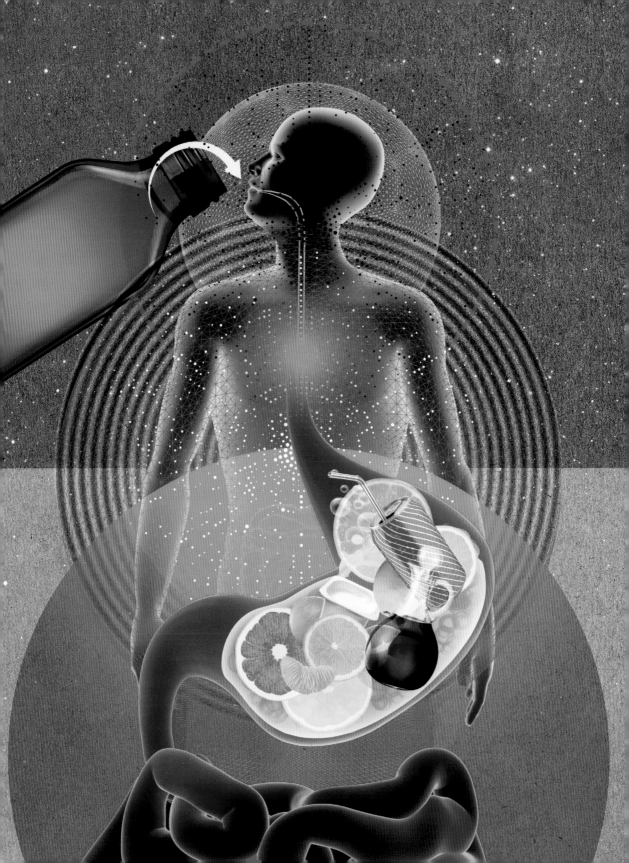

CREATING SOLIDS: PRECIPITATION REACTIONS

3-SECOND NUCLEUS

A precipitation reaction occurs when two ions are so strongly attracted to each other that they form a solid.

3-MINUTE VALENCE

Precipitation reactions are particularly useful when something needs to be removed from a solution. For example, water treatment plants can use precipitation reactions to remove undesirable contaminants from our water. If a chemical reaction takes place in a solution, and the product is insoluble in that solution, then it falls out of solution as a solid and can be isolated by filtration.

Two friends are out for a night on the town, both looking for love. One is drawn to an incredibly attractive stranger. Once together, the friend and the stranger never separate, because the attraction is so strong. They drop out of sight, losing touch with their old friends. This scenario is similar to a particular type of chemical reaction called a precipitation reaction. When two or more ions are mixed in water, they are initially attracted to nearby water molecules. However, as they move around (due to thermal energy), the ions encounter each other and are drawn together. If the attraction is strong enough, the two ions will stick together and fall out of solution (form a solid). If the attraction isn't very strong, the ions don't get together and just stay in solution. As examples, when silver (Ag^+) and chloride (Cl^-) are mixed they stick together and form a solid precipitate. However, sodium (Na^+) is much less attractive to chloride, so they don't form a precipitate. The "bathtub ring" that you sometimes see if you bathe in hard water is due to a precipitation reaction between the ions in hard water and the ions in soap.

RELATED TOPICS

See also
OPPOSITES ATTRACT
page 36

THE SOLID STATE
page 48

CERAMICS
page 52

3-SECOND BIOGRAPHY
LINUS PAULING
1901–94
American winner of the 1954 Nobel Prize in Chemistry, who developed our understanding of how atoms and molecules are attracted to each other; he was also a peace activist and won the 1962 Nobel Peace Prize

EXPERT
Jeff C. Bryan

In a precipitation, a solid forms when two liquid solutions are mixed.

USING CHEMISTRY TO GENERATE ELECTRICITY

3-SECOND NUCLEUS

Loss of electrons is oxidation, while gain of electrons is reduction; reactions that involve transfer of electrons between reactants are oxidation-reduction or redox reactions.

3-MINUTE VALENCE

Corrosion results from oxidation of metals exposed to oxidizing agents in the environment. When the metal is iron, the process is rusting. Rust is hydrated forms of iron(III) oxide generated when iron is exposed to moisture and oxygen. The rate of rusting depends on the acidity of the environment and the presence of electrolytes to help carry electric current. Having a metal that is more easily oxidized (a sacrificial electrode) in contact with iron can retard rusting.

Chemical reactions in which electrons migrate from one chemical substance to another are called oxidation-reduction reactions. These types of reactions can be used to generate electricity by arranging the chemical substances so that the substance gaining electrons (being reduced and called the oxidizing agent) is not in physical contact with the substance losing electrons (being oxidized and called the reducing agent). The electrons are then forced to travel through an external circuit to get from the reducing agent to the oxidizing agent. This arrangement is called a voltaic cell. Self-contained voltaic cells—either by themselves or connected in series—function as batteries. They produce electricity. The lead storage battery used to start a car engine is comprised of six voltaic cells containing lead and lead oxide immersed in a solution of sulfuric acid (battery acid). Dry cell batteries, such as those used in torches, use zinc and magnesium dioxide. Button batteries (used in calculators or watches) also use zinc, but they have mercuric oxide or silver oxide substituted as the oxidizing agent. Lithium ion batteries use lithium between planes of graphite as the reducing agent and a lithium metal oxide as the oxidizing agent.

RELATED TOPIC

See also
CHEMICAL EQUATIONS
page 60

3-SECOND BIOGRAPHIES

MICHAEL FARADAY
1791–1867
English scientist who developed the system of oxidation numbers and coined many terms associated with electrochemistry

WALTHER HERMANN NERNST
1864–1941
German chemist who developed the equation for the relationship between concentration and voltage

EXPERT
John B. Vincent

Batteries use chemical reactions that involve the transfer of electrons to produce electricity.

REACTION RATES & CHEMICAL KINETICS

3-SECOND NUCLEUS
The rate of a chemical reaction is the speed at which the reaction occurs and depends on reactant concentration, temperature, and whether or not a catalyst is present.

3-MINUTE VALENCE
Although nitrogen and oxygen gas are stable, at the temperature of a car engine they react to produce nitric oxide gas (NO), an air pollutant that is a cause of acid rain and smog. NO is removed from a car's exhaust gas by a catalytic converter. In the converter, the exhaust passes over a honeycomb-like structure of alumina impregnated with solid transition-metal oxides that catalyze the conversion of NO back to O_2 and N_2.

Chemical reactions occur at a variety of speeds or rates. Chemical explosions occur rapidly, with the creation of large volumes of hot gas. Many chemicals are stable, however; they react so slowly that they can be placed in a bottle. Fortunately, the rates (or speeds) at which chemical reactions occur can be controlled —and studying these processes is the field of chemical kinetics. One way to increase the rate of a reaction is by increasing the temperature. For this reason, the reactions that cook food happen faster the higher the temperature. Increasing the concentration of the reactant substances or surface area also increases the rates of chemical reactions. You can hold an iron nail in your hand, but if it is ground to a fine powder (greatly increasing its surface area) the iron can burst into flames in air. The concentration of many chemicals in your body must be carefully regulated for you to remain healthy. Your body accomplishes this by regulating the rates of a wide range of chemical reactions. The rates of slow reactions are accelerated by proteins called enzymes; these biological molecules are catalysts (substances that change the rate of a chemical reaction without being consumed by the reaction).

RELATED TOPICS
See also
CHEMICAL EQUATIONS
page 60

USING CHEMISTRY TO
GENERATE ELECTRICITY
page 68

3-SECOND BIOGRAPHIES
JACOBUS HENRICUS
VAN 'T HOFF
1852–1911
Dutch chemist who won the first Nobel Prize in Chemistry in part for determining graphical methods to establish that reaction rates depend on the concentrations of the reactants

HENRY TAUBE
1915–2005
American chemist who won the 1983 Nobel Prize in Chemistry for relating rates of chemical reactions to electronic structure

EXPERT
John B. Vincent

Controlling how fast a chemical reaction occurs allows us to reduce pollution and create new molecules.

1778
Born in Penzance, in southwestern England

1795
Apprenticed to a surgeon and apothecary

1797
Studies chemistry by reading in French the recently guillotined Antoine Lavoisier's *Traité élémentaire de chimie*

1798
Becomes the director of Thomas Beddoes's Pneumatic Institute in Bristol

1799
Prepares and imbibes nitrous oxide, which he nicknames "laughing gas"

1800
Publishes *Researches, Chemical and Philosophical; chiefly concerning nitrous oxide or dephlogisticated nitrous air, and its respiration*

1800
The news of Alessandro Volta's 1796 discovery of a chemical battery, known as the "Voltaic Pile," comes to England

1801
Invited to be a lecturer at the newly formed Royal Institution in London

1802
Appointed Professor of Chemistry at the Royal Institution; soon builds a giant Voltaic pile in the basement

1807-1808
Using his Voltaic pile, discovers six elements (sodium, potassium, magnesium, calcium, strontium, and barium) in two years

1810
"Discovers" and launches Michael Faraday on a stellar chemistry career

1812
Knighted, marries, retires, and with his new wife and Faraday, tours the continent visiting major laboratories

1815
Invents the coal miner's safety lamp

1829
After a lengthy period of debilitation, dies in Geneva, Switzerland

HUMPHRY DAVY

Humphry Davy, "full of mischief, with a penchant for explosions . . . a born chemist," spent his youth fishing, hunting, reading, storytelling, and writing poetry. As a teenager he loved fireworks and other explosive chemical reactions. He was apprenticed to a surgeon and apothecary, but reluctantly gave this up to become director of Thomas Beddoes's Pneumatic Institute in Bristol. This clinic, established to study the effect of gases on improving human health, gave Davy the opportunity to prepare, characterize, and purify nitrous oxide—which earlier researchers had thought caused the plague. Davy was not convinced and took a small whiff for himself. Noting that he did not meet his demise, he soon discovered that this sweet-smelling gas was thoroughly intoxicating. Fellow partakers giggled and laughed out loud, and Davy called it "laughing gas."

In Bristol, Davy became friends with a broad cross-section of intellectuals. His good looks, personal charm, and storytelling abilities, combined with his startling discoveries, made him a rising star. Although it soon became clear that Beddoes's clinic would be short-lived, Davy had matured into an excellent chemist. Luckily, he soon found an electrifying new field of study occasioned by Alessandro Volta's invention of a chemical battery, known then as a Voltaic pile. Characteristically, one of the first things he did was to build his own pile and shock himself to gauge its effectiveness.

In 1801, Davy was invited to be a lecturer at the newly minted Royal Institution in London. Here he quickly produced a series of immensely popular lectures augmented by striking demonstrations. He was an extremely handsome young man with a gift for working audiences, so he attracted huge crowds that included many of the young ladies of that day. One such admirer admitted that "those eyes were made for something besides poring over crucibles." Active in the electrolysis of metal oxides, he discovered six elements (sodium, potassium, magnesium, calcium, strontium, and barium) in two years (1807–08) and became the leading chemist of his day.

His penchant for self-experimentation caused debilitating injuries that resulted in his hiring of Michael Faraday as his assistant. Davy often claimed that Faraday was his most important "discovery." In 1812, Davy was knighted, married, and embarked on a Grand Tour of Europe. Upon his return he invented the coal miner's safety lamp that saved many lives. He was essentially an invalid for the last third of his life and died shortly after his fiftieth birthday. Nevertheless, during his short and exciting lifetime, he was one of the best practitioners of and spokesmen for the new science of chemistry.

Glen E. Rodgers

ENERGY & THE FIRST LAW OF THERMODYNAMICS

The universe has a quantity we call energy. Objects can possess it and can transfer it to other objects, but it can neither be created nor destroyed. The total amount of energy that exists is constant. This principle is known as the first law of thermodynamics. (A more nuanced treatment includes mass/energy as the constant, but we simplify a bit here.) We formally define energy as the capacity that an object has to exert a force on another object across a distance. For example, a moving car has energy because it has the capacity to strike another object and exert a force on it over a distance. Energy can come in many different forms. The moving car has kinetic energy, the energy associated with its motion. All substances above zero kelvin have thermal energy, a type of kinetic energy associated with the random, temperature-dependent motion of the particles that compose the substance. The higher the temperature, the greater the thermal energy. The book you are holding contains potential energy, the energy associated with an object's position. Chemical substances have chemical energy, a type of potential energy associated with the positions of all of their electrons and protons. Energy can be transferred or exchanged but, according to the first law, it can never be created or destroyed.

3-SECOND BIOGRAPHY

RUDOLF CLAUSIUS
1822–88
German physicist who formulated one of the earliest versions of the first law of thermodynamics

EXPERT

Nivaldo Tro

A steam engine is powered by the energy transferred from burning fuel.

ENTROPY & THE SECOND LAW OF THERMODYNAMICS

3-SECOND NUCLEUS

For all spontaneous processes, entropy increases.

3-MINUTE VALENCE

The second law of thermodynamics implies that a perpetual motion machine—one that keeps on moving forever without the need for energy input—is impossible. With each cycle of the machine's motion, some energy must be dispersed in order for the motion to occur at all. As a result, the energy of the machine must necessarily decrease over time, and it must eventually stop moving.

We have seen that in energy transactions, you can't win—you can't get energy out of nothing. But it gets worse—you can't even break even. In our universe, energy always spreads out or randomizes itself as much as possible. The second law of thermodynamics describes this pervasive tendency: in any spontaneous process, a quantity called entropy (which you can think of as a measure of energy randomization or energy dispersion) always increases. You have no doubt experienced the second law every time you hold a warm drink. The thermal energy in the drink disperses itself into the surroundings—the drink spontaneously cools down (and the air surrounding the drink warms up a bit). Imagine a universe in which the hot drink gets hotter (and the surroundings slightly cooler) as energy transfers from the surroundings into the drink. Not possible according to the second law. The second law implies that, in any energy transaction, some energy must be dispersed if the transaction is to occur at all. In other words, nature always takes a heat tax. For example, recharging a battery will always take more energy than the amount of energy you can use as you discharge the battery. Such is the second law—when it comes to energy, you can't break even.

RELATED TOPICS

See also
ENERGY & THE FIRST LAW OF THERMODYNAMICS
page 74

ENTROPY & THE THIRD LAW OF THERMODYNAMICS
page 78

ENTROPY & SPONTANEOUS PROCESSES
page 80

3-SECOND BIOGRAPHIES

NICOLAS LÉONARD SADI CARNOT
1796–1832
French physicist instrumental in the development of thermodynamics

RUDOLF CLAUSIUS
1822–88
German physicist who was instrumental in formulating the second law of thermodynamics

EXPERT

Nivaldo Tro

A perpetual motion machine cannot exist according to the second law of thermodynamics.

ENTROPY & THE THIRD LAW OF THERMODYNAMICS

3-SECOND NUCLEUS

The absolute entropy of a perfect crystal at zero kelvin is zero.

3-MINUTE VALENCE

Entropy is a measure of the energy dispersed into a system per unit temperature. For this reason, dispersing the same amount of energy into a system at a colder temperature produces greater entropy than dispersing that energy into a warmer system. As we will see in the next entry, this is the reason that ice melts above its melting point but not below it.

According to the first law of thermodynamics, you can't win—you can't get something for nothing. According to the second law, you can't break even—every energy transaction necessarily results in a loss to the surroundings. According to the third law of thermodynamics, you can't get out of the game. In the case of thermodynamics "getting out of the game" means getting to the lowest possible temperature, zero kelvin or absolute zero. Absolute zero is the temperature at which atomic and molecular motion essentially stops. The third law of thermodynamics states that the entropy of a perfect crystal is zero at zero kelvin. This law has two implications. The first one is that entropy, unlike other thermodynamic quantities, can be measured on an absolute scale. All perfect crystals have zero entropy at zero kelvin. As the temperature rises, energy is dispersed into the crystal and its temperature and entropy increases. The second implication is that absolute zero can never be reached in a finite number of steps. It would take an infinite number of cooling steps to arrive at the absolute zero of temperature, so it can never be achieved.

RELATED TOPICS

See also
ENERGY & THE FIRST LAW OF THERMODYNAMICS
page 74

ENTROPY & THE SECOND LAW OF THERMODYNAMICS
page 76

ENTROPY & SPONTANEOUS PROCESSES
page 80

3-SECOND BIOGRAPHY

WALTHER HERMANN NERNST
1864–1941
German chemist who formulated the third law of thermodynamics and received the 1920 Nobel Prize in Chemistry for his work

EXPERT

Nivaldo Tro

The third law of thermodynamics, as formulated by Walther Nernst, implies that the absolute zero of temperature cannot be reached.

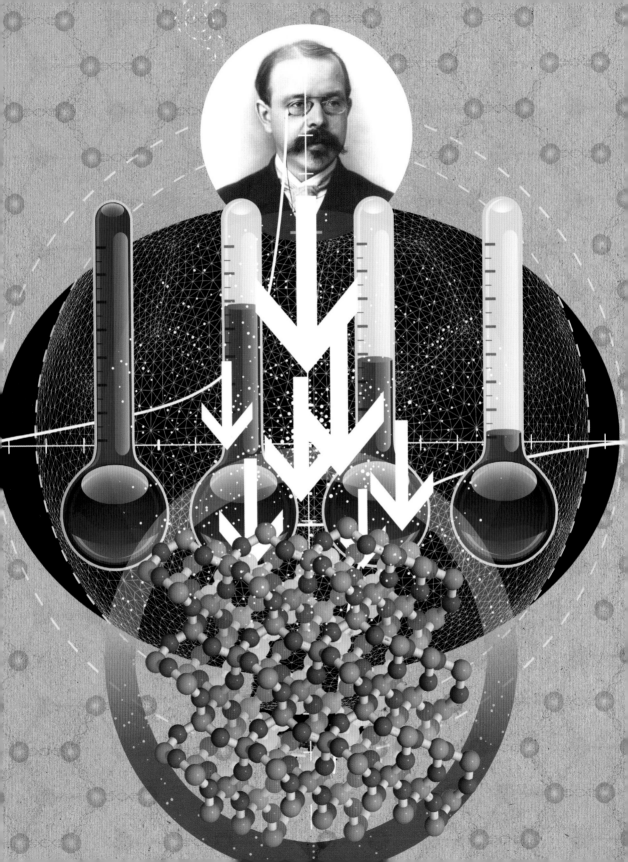

ENTROPY & SPONTANEOUS PROCESSES

3-SECOND NUCLEUS
A process is spontaneous if it increases the entropy (energy dispersion) of the universe.

3-MINUTE VALENCE
Processes that result in a decrease in entropy are not impossible, they just don't happen spontaneously. Iron spontaneously reacts with oxygen to form iron oxide (rust). This process causes an increase in the entropy of the universe. However, iron oxide can be turned back into iron. In fact, the manufacture of iron metal from iron oxide depends on it.

The criteria for determining whether any process will happen is simple: will the process result in an increase in entropy? Consider the freezing of water. Water freezes spontaneously below 32°F (0°C). Why? When water freezes, the water molecules become more organized, and the energy they contain becomes less randomized—their entropy decreases. How then is this process ever spontaneous? Because when water freezes it gives off heat (energy is dispersed) and the resulting entropy increase in the surroundings is temperature-dependent. We can understand this with a simple analogy. If you give a poor person $1,000, you significantly increase his or her net worth. But if you give a rich person the same $1,000, the impact is negligible. Similarly, if you disperse a given amount of energy into cold surroundings, you significantly increase its entropy, but if you disperse the same amount of energy into warm surroundings, the increase is less. When the freezing of water occurs below 32°F, the heat emitted into the surroundings is enough to increase the entropy of the surroundings so much that it more than compensates for the decrease in entropy of the water molecules themselves, resulting in an overall increase in entropy of the universe and therefore a spontaneous process.

RELATED TOPICS
See also
ENERGY & THE FIRST LAW OF THERMODYNAMICS
page 74

ENTROPY & THE SECOND LAW OF THERMODYNAMICS
page 76

ENTROPY & THE THIRD LAW OF THERMODYNAMICS
page 78

3-SECOND BIOGRAPHIES
JOSIAH WILLARD GIBBS
1839–1903
American physicist who developed the main criteria for the spontaneity of a process

LUDWIG EDUARD BOLTZMANN
1844–1906
Austrian physicist who developed a statistical description of the second law of thermodynamics

EXPERT
Nivaldo Tro

When ice melts at a temperature above its melting point, the entropy of the universe increases.

INORGANIC CHEMISTRY

allotropes Two or more forms of the same element, but with different structures (and therefore different properties).

catalytic properties To have the ability to act as a catalyst (a substance that increases the rate of chemical reaction without being consumed by the reaction).

CFCs Chlorofluorocarbons. These compounds were common in air conditioning and refrigeration, but are now banned due to their harmful effect on the Earth's ozone layer.

color wheel A circle or wheel that contains different colors and shows how they are related. You can use a color wheel to predict the color of an object based on what colors the object absorbs.

complementary colors Two colors opposite each other on a color wheel. Complementary colors have high contrasts between one another.

complex (transition metal complex) A compound or ion consisting of a transition metal linked to one or more ligands.

concatenated atoms Atoms that have been linked to form a chain structure.

crystal field theory (also ligand field theory) A bonding theory in inorganic chemistry in which ligands donate an electron pair to a central metal ion.

diffraction grating A surface engraved with a series of closely spaced lines that reflects different wavelengths of light at different angles. Diffraction gratings can split white light up into its constituent colors.

electromagnetic spectrum The range of frequencies of electromagnetic radiation bounded by radio waves at low frequencies and gamma rays at high frequencies.

fuel cell An electrochemical cell that produces electrical current from the continuous input of a fuel.

fullerenes Carbon molecules that have spherical, tubular, or other similar structures.

graphene A form of carbon consisting of a sheet of carbon atoms one atom thick.

graphite A form of carbon composed of carbon atoms bound together in sheets, which are stacked on top of one another.

ligand A molecule or ion that donates an electron pair to a central metal ion in transition metal complex.

metalloid An element that falls along the boundary between metals and nonmetals on the periodic table. Metalloids have properties intermediate between metals and nonmetals.

oxidation state The charge an atom would have in a chemical compound if all of the bonding electrons were assigned to the more electron-negative atom (the atom that most strongly attracts electrons).

photosynthesis The process by which plants convert carbon dioxide, water, and sunlight into glucose and oxygen.

prism A clear optical element that is usually triangular in shape and can bend light of different wavelengths by different amounts. When white light travels through a prism, it is broken up into its constituent colors.

reactant Any one of the substances that undergoes a chemical reaction. In a reaction, reactants react to form products.

silicate Compound containing silicon, oxygen, and sometimes various metal atoms. Silicates form network covalent structures with high melting points.

stratosphere An atmospheric layer that begins about 6 miles (10 km) above the Earth's surface and is sandwiched between the troposphere below and the mesosphere above.

substrate The molecule on which an enzyme (a biological catalyst) acts.

transition metals Those metals found in the large center block of the periodic table (the d-block). Transition metals (in contrast to main group metals) tend to have properties that are less predictable based on their exact position on the periodic table.

valence electrons The highest-energy electrons (and therefore the most important in bonding) in an atom.

Zintl ions Ionic clusters of main group elements.

THE UNIQUENESS PRINCIPLE

The periodic table contains a

dividing line that is never marked. The invisible line occurs between the table's second and third rows, where boron meets aluminium; carbon meets silicon; and so on until fluorine meets chlorine. Elements above the line cannot form bonds to as many atoms as those below the line, being strictly limited to a total of eight "valence" or bonding-level electrons. Thus, while oxygen forms the mono- and dioxides with itself (that is, O_2 and O_3), sulfur, selenium, and tellurium form mono-, di-, and trioxides, of which SO, SO_2, and SO_3 are examples. Similarly, nitrogen forms the trichloride NCl_3 while phosphorous, arsenic, and antimony form both tri- and pentachlorides such as PCl_3 and PCl_5. However, unlike later elements, the nonmetals of the second row are small enough to form strong multiple bonds, a quality that enables them to form compact molecules where a heavier element would form an extended structure of linked atoms. Carbon, for instance, forms the triple- and doubly bonded oxides CO and CO_2, while its heavier counterparts—silicon, germanium, tin, and lead—react with oxygen to form three-dimensional solid networks held together exclusively by single bonds.

3-SECOND NUCLEUS
Elements in the periodic table's second row behave differently from heavier elements because they form strong multiple bonds and are limited to eight valence electrons.

3-MINUTE VALENCE
One consequence of the uniqueness principle is that planetary atmospheres largely consist of first- and second-row elements. Earth's atmosphere is comprised mostly of nitrogen and oxygen; Mars's atmosphere mostly of CO_2; and those of the gas giants are mostly hydrogen, helium, CH_4, and NH_3. In contrast, the planetary crusts of planets such as Earth and Mars contain large amounts of silicate minerals, many of which contain chains, sheets, and 3-D networks held together by silicon-oxygen single bonds.

RELATED TOPICS
See also
PERIODIC PATTERNS
page 26

THE LEWIS MODEL FOR
CHEMICAL BONDING
page 32

CARBON: IT'S NOT JUST
FOR PENCILS
page 96

3-SECOND BIOGRAPHIES
VICTOR GOLDSCHMIDT
1888–1947
Swiss crystal chemist who classified elements by their dominant geologic locations

THOM DUNNING
1943–
American chemist who explained first-row anomalies in terms of recoupled pair bonds

EXPERT
Stephen Contakes

The second-row elements are unique— they are very different from the elements that lie below them in the periodic table.

1 H				
H				
Hydrogen				
1.00794				

2 He
He
Helium
4.002602

3 Li	4 Be
Li	Be
Lithium	Beryllium
6.941	9.012182

5 B	6 C	7 N	8 O	9 F	10 Ne
B	C	N	O	F	Ne
Boron	Carbon	Nitrogen	Oxygen	Fluorine	Neon
10.811	12.0107	14.0067	1.00794	18.9984032	20.1797

11 Na	12 Mg												13 Al	14 Si	15 P	16 S	17 Cl	18 Ar
Na	Mg												Al	Si	P	S	Cl	Ar
Sodium	Magnesium												Aluminium	Silicon	Phosphorus	Sulfur	Chlorine	Argon
22.98976928	24.305												26.9815386	28.0855	30.973762	32.065	35.453	39.948
19 K	20 Ca	21 Sc	22 Ti	23 V	24 Cr	25 Mn	26 Fe	27 Co	28 Ni	29 Cu	30 Zn	31 Ga	32 Ge	33 As	34 Se	35 Br	36 Kr
K	Ca	Sc	Ti	V	Cr	Mn	Fe	Co	Ni	Cu	Zn	Ga	Ge	As	Se	Br	Kr
Potassium	Calcium	Scandium	Titanium	Vanadium	Chromium	Manganese	Iron	Cobalt	Nickel	Copper	Zinc	Gallium	Germanium	Arsenic	Selenium	Bromine	Krypton
39.0983	40.078	44.955912	47.867	50.9415	51.9961	54.938045	55.845	58.933195	58.6934	63.546	65.38	69.723	72.64	74.92160	78.96	79.904	83.798
37 Rb	38 Sr	39 Y	40 Zr	41 Nb	42 Mo	43 Tc	44 Ru	45 Rh	46 Pd	47 Ag	48 Cd	49 In	50 Sn	51 Sb	52 Te	53 I	54 Xe
Rb	Sr	Y	Zr	Nb	Mo	Tc	Ru	Rh	Pd	Ag	Cd	In	Sn	Sb	Te	I	Xe
Rubidium	Strontium	Yttrium	Zirconium	Niobium	Molybdenum	Technetium	Ruthenium	Rhodium	Palladium	Silver	Cadmium	Indium	Tin	Antimony	Tellurium	Iodine	Xenon
85.4678	87.62	88.90585	91.224	92.90638	95.96	[98]	101.07	102.90550	106.42	107.8682	112.411	114.818	118.71	121.76	127.6	126.90447	131.293
55 Cs	56 Ba	57-71 Lanthanoids	72 Hf	73 Ta	74 W	75 Re	76 Os	77 Ir	78 Pt	79 Au	80 Hg	81 Tl	82 Pb	83 Bi	84 Po	85 At	86 Rn
Cs	Ba		Hf	Ta	W	Re	Os	Ir	Pt	Au	Hg	Tl	Pb	Bi	Po	At	Rn
Caesium	Barium		Hafnium	Tantalum	Tungsten	Rhenium	Osmium	Iridium	Platinum	Gold	Mercury	Thallium	Lead	Bismuth	Polonium	Astatine	Radon
132.9054519	137.327		178.49	180.94788	183.84	186.207	190.23	192.217	195.084	196.966569	200.59	204.3833	207.2	208.98040	[209]	[210]	[222]
87 Fr	88 Ra	89-103 Actinoids	104 Rf	105 Db	106 Sg	107 Bh	108 Hs	109 Mt	110 Ds	111 Rg	112 Cn	113 Nh	114 Fl	115 Mc	116 Lv	117 Ts	118 Og
Fr	Ra		Rf	Db	Sg	Bh	Hs	Mt	Ds	Rg	Cn	Nh	Fl	Mc	Lv	Ts	Og
Francium	Radium		Rutherfordium	Dubnium	Seaborgium	Bohrium	Hassium	Meitnerium	Darmstadtium	Roentgenium	Copernicium	Nihonium	Flerovium	Moscovium	Livermorium	Tennessine	Oganesson
[223]	[226]		[267]	[268]	[271]	[272]	[270]	[276]	[281]	[280]	[285]	[284]	[289]	[288]	[293]	[294]	[294]

57 La	58 Ce	59 Pr	60 Nd	61 Pm	62 Sm	63 Eu	64 Gd	65 Tb	66 Dy	67 Ho	68 Er	69 Tm	70 Yb	71 Lu
La	Ce	Pr	Nd	Pm	Sm	Eu	Gd	Tb	Dy	Ho	Er	Tm	Yb	Lu
Lanthanum	Cerium	Praseodymium	Neodymium	Promethium	Samarium	Europium	Gadolinium	Terbium	Dysprosium	Holmium	Erbium	Thulium	Ytterbium	Lutetium
138.90547	140.116	140.90765	144.242	[145]	150.36	151.964	157.25	158.92535	162.5	164.93032	167.259	168.93421	173.054	174.9668
89 Ac	90 Th	91 Pa	92 U	93 Np	94 Pu	95 Am	96 Cm	97 Bk	98 Cf	99 Es	100 Fm	101 Md	102 No	103 Lr
Ac	Th	Pa	U	Np	Pu	Am	Cm	Bk	Cf	Es	Fm	Md	No	Lr
Actinium	Thorium	Protactinium	Uranium	Neptunium	Plutonium	Americium	Curium	Berkelium	Californium	Einsteinium	Fermium	Mendelevium	Nobelium	Lawrencium
[227]	232.03806	140.90765	238.02891	[237]	[244]	[243]	[247]	[247]	[251]	[252]	[258]	[258]	[259]	[262]

COLOR

RELATED TOPICS
See also
CLUSTER CHEMISTRY
page 90

TRANSITION METAL
CATALYSTS
page 92

When white light passes through a prism, the light is dispersed into a spectrum of colors. The colors range from red at the lowest frequencies through orange, yellow, green, blue, and finally violet at the highest. Substances that absorb all frequencies of visible light appear black, while substances that reflect all frequencies of visible light appear white. Colored objects appear to an observer to have a color because they absorb certain frequencies (or wavelengths) of visible light while reflecting or transmitting (allowing to pass) others. The precise color a substance has depends on which frequencies are absorbed. In general, a substance will appear to have a color complementary to the one absorbed (and opposite it on a color wheel). For example, a substance that appears yellow absorbs violet light (the complement of yellow). Transition metal complexes are often deeply colored because they strongly absorb certain frequencies of light in the visible region. These complexes often have unfilled outer-level orbitals that can receive an electron excited by specific frequencies of visible light. The color absorbed depends on the separations between the d-orbitals, which in turn depend on the ligands attached to the metal.

3-SECOND NUCLEUS
Colored objects appear to have color because they absorb some frequencies of visible light and reflect or transmit the others.

3-MINUTE VALENCE
Our eyes can detect a narrow range of frequencies in the electromagnetic spectrum. This range of frequencies is responsible for all of the colors that we see. Our brains have evolved to use color as a way to help distinguish one substance from another. Modern spectrometers, which precisely measure the frequencies absorbed by substances, are among the most powerful scientific tools in substance identification.

3-SECOND BIOGRAPHIES
ALFRED WERNER
1866–1919
Swiss chemist who won the 1913 Nobel Prize in Chemistry for predicting the three-dimensional structure of many transition metal complexes before modern structural methods

JOHN HASBROUCK VAN VLECK
1899–1980
American physicist who won the 1977 Nobel Prize in Physics and was instrumental in the development of crystal field theory, the precursor to ligand field theory

EXPERT
John B. Vincent

White light separates into its component colors when passed through a prism.

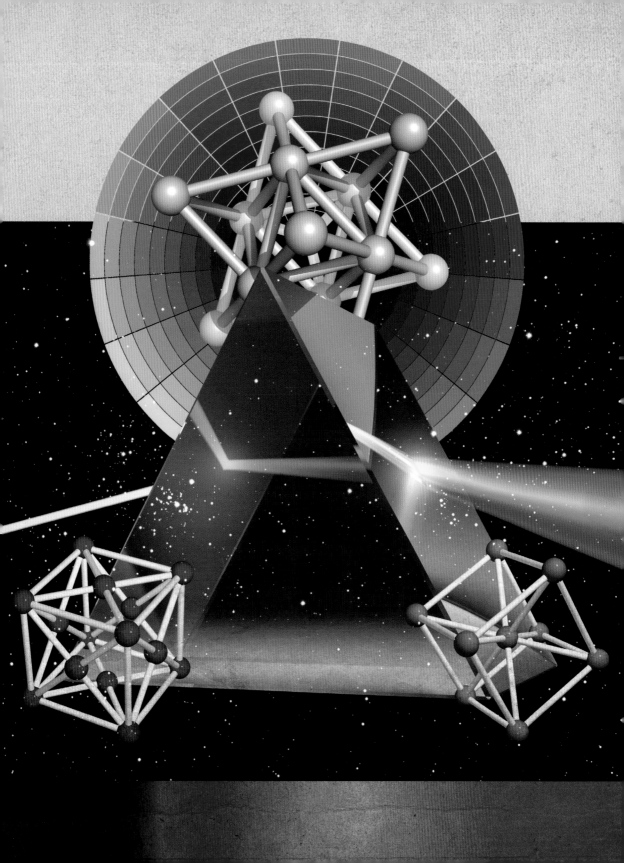

CLUSTER CHEMISTRY

In some molecules, ions, and materials, the electrons that hold the atoms together are shared between a group of clustered atoms. For instance, some metals and metalloids can be reduced to liquid-soluble fragments of metal while many transition metals form clusters when combined with chlorine, sulfur, or carbon monoxide under the right conditions. Some of the latter clusters catalyze commercially important reactions, although none has yet found industrial use. Sometimes clusters exist as discrete units; in other cases they are linked together in a network. Solid $MoCl_2$, for example, consists of octahedral clusters of $Mo_6Cl_8^{4+}$ bridged by intervening Cl^- ions. However, when heated in the presence of additional chloride, the connections are broken to give discrete $Mo_6Cl_{14}^{2-}$ units. The ratio of electrons to atoms in a cluster affects its shape. When clusters have just enough electrons to hold together, they form as compact a shape as possible—that of the smallest polyhedron that can accommodate all the core atoms. In contrast, clusters with more electrons tend to open up and take on the shape of larger polyhedra with unoccupied vertices, giving clusters that sometimes look like a molecular nest or a web.

3-SECOND NUCLEUS

Cluster compounds form when atoms and ions share electrons and bunch together in a polyhedral shape.

3-MINUTE VALENCE

Many of the most important reactions for life on Earth are facilitated by metal-containing clusters located within proteins. For example, clusters containing iron and sulfur facilitate the movement of electrons through many biological systems, including the respiratory chain our cells use to harvest energy by converting oxygen to water. In photosynthesis plants reverse this process, harvesting even more energy from sunlight to produce oxygen at a cluster containing four manganese ions and one calcium ion.

RELATED TOPICS

See also
WHERE ELECTRONS ARE WITHIN AN ATOM
page 24

THE LEWIS MODEL FOR CHEMICAL BONDING
page 32

NANOTECHNOLOGY
page 98

3-SECOND BIOGRAPHIES

WILLIAM N. LIPSCOMB
1919–2011
American chemist who pioneered study of the structure and bonding in borane clusters

KENNETH WADE
1932–2014
British chemist who developed "Wade's Rules" for predicting cluster compounds' shapes and stability

EXPERT

Stephen Contakes

Cluster compounds have polyhedral shapes and unique properties.

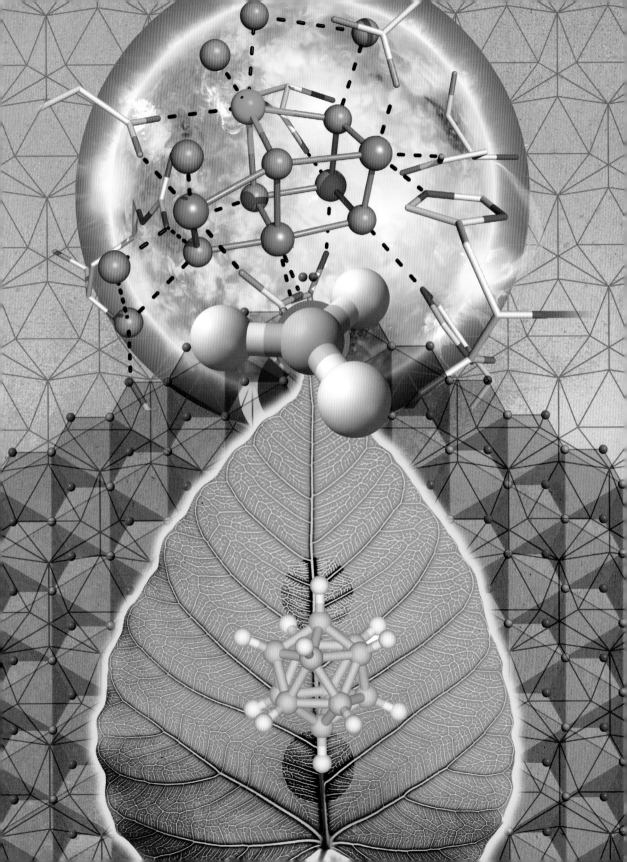

TRANSITION METAL CATALYSTS

Many industrial chemicals are produced by combining small organic molecules or substrates with catalysts composed of transition metals (such as cobalt, chromium, or iron), which are themselves bound to small molecules called ligands. In these processes the metals speed up reactions between the substrates by acting as platforms where substrates can bind and then break into smaller molecular fragments, rearrange how their atoms are bound together, and form new bonds with other substrates. The new molecules and fragments that result can then be released from the metal to yield the reaction products and regenerate the original metal complex. In fact, from the viewpoint of the metal the entire process involves a cyclic series of reactions or "catalytic cycle" in which the metal complex adds reactants and spits out products. Some metal catalysts do not even need to bind their substrates, but instead function by pushing electrons around. For instance, some biological iron clusters facilitate the movement of electrons between reactants, alternatively gaining an electron from one substrate and passing it to another. Other complexes can harvest energy from light and use it to push electrons into or out of molecules, generating unstable intermediates that then quickly react with other nearby molecules.

3-SECOND NUCLEUS
Transition metal compounds facilitate chemical reactions between small molecules that bind to the metal, rearrange, and get released as new products.

3-MINUTE VALENCE
Many biological and industrial processes involve transition metals acting as catalysts, substances that speed up chemical reactions without being changed themselves. Without metal catalysts we could not use the oxygen we breathe or produce enough food to sustain current population levels. Perhaps we wouldn't even be around today, because one hypothesis about the origin of life involves catalysis by iron minerals.

RELATED TOPICS
See also
REACTION RATES & CHEMICAL KINETICS
page 70

COLOR
page 88

AMINO ACIDS & PROTEINS
page 128

3-SECOND BIOGRAPHIES
HUMPHRY DAVY
1779–1829
British chemist who discovered that platinum was useful as a heterogeneous catalyst

KARL ZIEGLER
& GIULIO NATTA
1898–1973 & 1903–79
German and Italian chemists who developed a catalyst for making commercial plastics with specific properties

EXPERT
Stephen Contakes

Transition metal compounds can act as catalysts in chemical reactions.

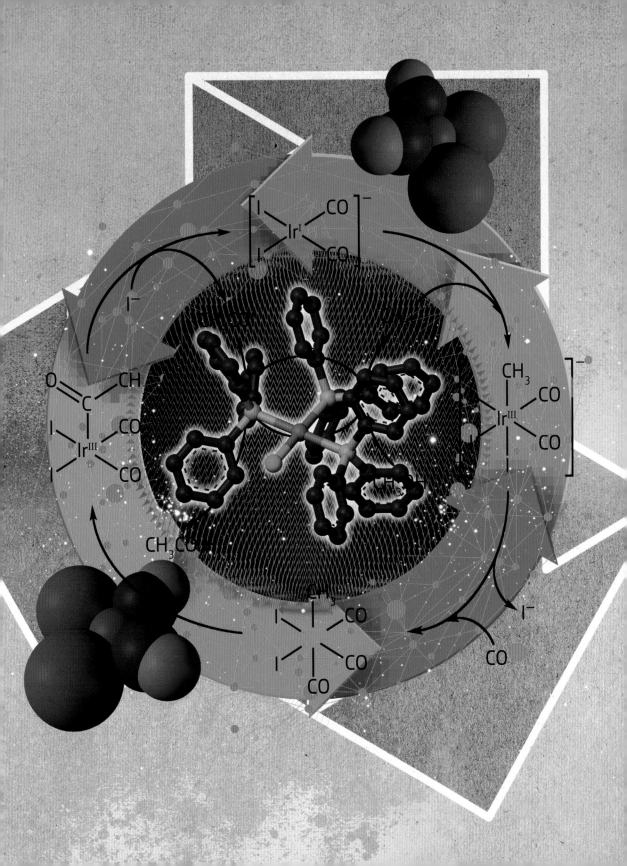

19 March 1943
Born in Mexico City, Mexico

1972
Receives his PhD in chemistry from the University of California, Berkeley

1973
Joins Professor Rowland's research group at the University of California, Irvine

1974
Coauthors a paper in the journal *Nature* highlighting the damaging effects of CFCs to the ozone layer in the stratosphere

1982
Moves to the Jet Propulsion Laboratory at the California Institute of Technology to carry out experiments on the effect of CFCs on the ozone layer

1985
Demonstrates that ice crystals in the polar stratosphere amplify the ozone destruction capability of CFCs

1989
Moves to Massachusetts Institute of Technology to continue his research in atmospheric sciences

1995
Receives the Nobel Prize in Chemistry for "contributing to our salvation from a potential global environmental catastrophe"

2005
Moves to the University of California, San Diego, and the Center of Atmospheric Sciences at Scripps Institution of Oceanography

2013
Receives the Presidential Award of Freedom from President Obama

MARIO J. MOLINA

Born to a Mexican family of

highly educated professionals, Mario J. Molina is a Nobel Prize–winning chemist whose work on the detrimental effects of chlorofluorocarbons (CFCs)—a class of industrial chemicals commonly used in refrigeration, aerosol cans, and plastic manufacturing—on the stratospheric ozone layer provides an excellent example of how fundamental research can have tremendous practical implications, greatly improving the quality of life on Earth. His research findings, published in the journal *Nature* in 1974, led to an international ban on CFC emissions into the atmosphere (the 1985 Vienna Convention and the Montreal Amendment).

Encouraged by his aunt Esther Molina, also a chemist, Mario developed a strong fascination with natural sciences, which became stronger when he observed the living organisms in a drop of pond water with his first microscope. This thrilling experience drew him to acquire chemistry sets and to build his own laboratory in an unused bathroom of his family home. Recognizing how excited he was about chemistry, Molina's parents briefly sent him to a Swiss boarding school at the age of eleven to learn German, a language then quite useful for a successful career in chemistry.

After studying chemical engineering at the Autonomous National University of Mexico, Molina spent more than two years in Germany and France strengthening his engineering and mathematics knowledge before completing his doctoral degree under the supervision of Professor George Pimentel at the University of California, Berkeley.

In 1973, Molina joined Professor Rowland's group as a postdoctoral researcher at the University of California, Irvine, where he discovered that CFCs break down in the stratosphere, producing elemental chlorine, which destroys the ozone layer that protects living things on Earth from the Sun's dangerous rays. Molina and Rowland met with fierce opposition from industrial producers of CFCs until the British Antarctic Survey detected a large and growing gap in the ozone layer in 1985. Molina, now a full-time researcher at the Jet Propulsion Laboratory at the California Institute of Technology, further demonstrated that polar ice crystals in the stratosphere amplified the destructive effect of CFCs on the ozone layer. By the end of 1985, most of the CFC-producing countries signed the Vienna Convention, which was soon amended by the Montreal Protocol, to end CFC emissions into the atmosphere. Mario Molina received the 1995 Nobel Prize in Chemistry for "contributing to our salvation from a potential global environmental catastrophe."

Ali O. Sezer

CARBON: IT'S NOT JUST FOR PENCILS

3-SECOND NUCLEUS

Carbon allotropes include diamond, graphite, and fullerenes. Fullerenes, named after Buckminster Fuller, come in the form of hollow spheroids ("buckyballs") and nanotubes ("buckytubes").

3-MINUTE VALENCE

Carbon is a unique element. Not only is it the central element on which all life is based, but in its elemental state it also exists in several fascinating and useful allotropes (different molecular forms of a given element).

The major common allotropes of carbon include shiny, transparent, and super-hard diamond and soft black graphite. Graphene (isolated in 2010 by Andre Geim) is a one-atom-thick sheet of carbon atoms densely packed into a chicken-wire-shaped structure. Graphite is many sheets of these graphene layers stacked together. Since these layers are only weakly bound to each other, they are readily rubbed off onto paper and therefore used in pencils. In 1985, Harry Kroto discovered a new form of carbon (C_{60}) when he and his colleagues Richard Smalley and Robert Curl simulated chemical reactions that might occur in the atmosphere of red giant stars. C_{60} is one of a larger family of carbon forms known as fullerenes. Fullerenes are molecules in the form of hollow spheres, ellipsoids, or cylinders. Since C_{60} looks much like the famous geodesic domes of R. Buckminster Fuller, it was dubbed "buckminsterfullerene." Spherical or ellipsoidal fullerenes are fondly known as "buckyballs." Graphene layers can also wrap around and form cylindrical tubes called "nanotubes" or "buckytubes." Fullerenes are flexible, strong, and stable, with an ever-increasing number of practical uses—for example, as catalysts, in energy generation and storage devices, as MRI and X-ray contrast agents, and in flexible electronics.

RELATED TOPICS

See also
BONDING ATOMS TOGETHER
page 28

THE LEWIS MODEL FOR
CHEMICAL BONDING
page 32

THE FORCES THAT HOLD
MATTER TOGETHER
page 42

3-SECOND BIOGRAPHIES

R. BUCKMINSTER FULLER
1895–1983
American inventor, architect, and author who popularized the geodesic dome

HARRY KROTO
1939–2016
English chemist, winner of the 1996 Nobel Prize in Chemistry for discovering fullerenes

EXPERT

Glen E. Rodgers

Carbon comes in many forms, including familiar graphite and new forms such as graphene.

NANOTECHNOLOGY

3-SECOND NUCLEUS
Nanotechnology is the
science of building tiny
structures from the
molecular size up.

3-MINUTE VALENCE
Nanochemistry can
also be used to construct
nanoparticles of precise
size and shape with unique
catalytic properties for
processes such as
converting hydrogen and
oxygen to water in a fuel
cell to generate electricity.
Other nanoparticles can be
used to split water, using
sunlight, to the requisite
hydrogen and oxygen
needed by the fuel
cell. Combined, such a
system could potentially
supply the world's energy
needs in a far cleaner
manner than using fossil
fuels—as we do today.

Nanotechnology is the study of matter where the basic structure has at least one of its dimensions less than or equal to 100 nanometers (1 nanometer is one-billionth of a meter). This would therefore include almost all of chemistry, since most molecules meet such a requirement. However, what is different about nanochemistry is that it uses bottom-up molecular routes to reach the larger domains, coupling molecule to molecule to make larger structures. This is in contrast to the traditional top-down approaches in which large structures are cut into smaller pieces. The bottom-up route permits the precision of chemical synthesis to affect larger materials properties. As an example of this bottom-up approach, different molecules that are about 0.1 nanometer in size can be attached together using synthetic chemical techniques to make tiny structures such as nanocars. A single nanocar is 2 nm x 3 nm in size with four wheels, fully rotating axles, chassis, and light-activated motors. These nanocars may be able to perform work, such as bringing in molecules or atoms through selective voltage pulse commands to further construct larger entities or to deliver drugs to cells. Molecularly built nanocars are so small that 25,000 of them lined up end to end would only span a distance the diameter of a human hair.

RELATED TOPICS
See also
CLUSTER CHEMISTRY
page 90

CARBON: IT'S NOT JUST
FOR PENCILS
page 96

3-SECOND BIOGRAPHIES
RICHARD FEYNMAN
1918–88
American physicist who first
suggested building molecular
structures one atom at a time;
he won the 1965 Nobel Prize in
Physics

RICHARD SMALLEY
1943–2005
American chemist and
foundational figure in
nanotechnology who was
awarded a share of the 1996
Nobel Prize in Chemistry

EXPERT
James Tour

*Molecular machines
such as nanocars
have the potential
to perform vital tasks
in medicine and
energy supply.*

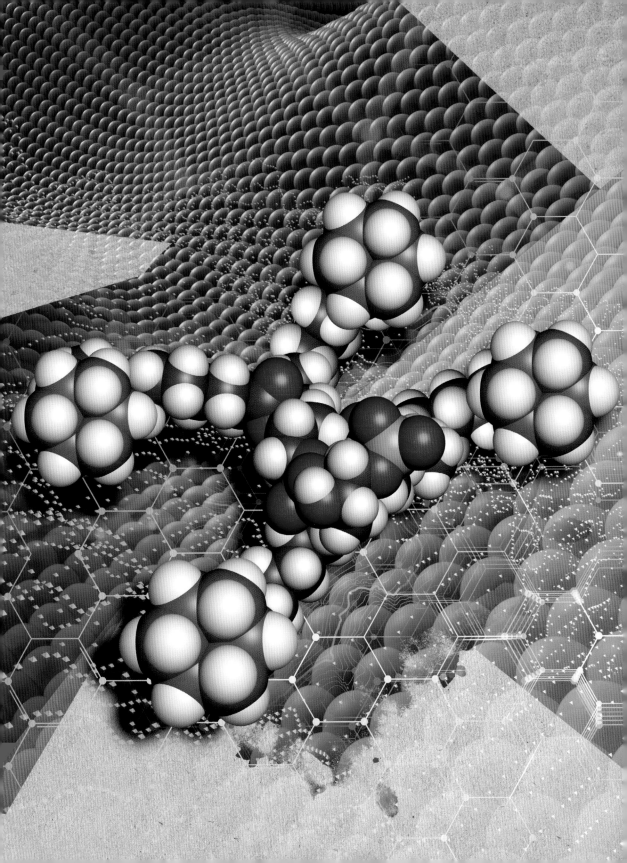

ORGANIC CHEMISTRY

aldehydes A family of organic compounds containing a terminal C=O functional group.

aldosterone A hormone responsible for the regulation of sodium and potassium.

alkaloids Organic bases found in plants; often poisonous.

amines Organic compounds containing a nitrogen atom bonded to one or more carbon atoms.

aromatic ring A flat, ring-shaped chain of carbon atoms containing alternating single and double bonds.

carbonyl The C=O functional group.

carboxylic acid An organic compound containing the –COOH functional group.

covalent bond The joining of atoms by the sharing of one or more electrons.

cross-linked structure A common structure in polymers in which long chainlike molecules form bonds between the neighboring chains.

cyclic structures Any structure containing one or more rings of atoms bonded together.

distillation A process by which a mixture of compounds with differing boiling points can be separated by heating the mixture and recondensing the gases that vaporize.

ester An organic compound containing the –COO– functional group.

functional group A characteristic atom or group of atoms present in a family of organic compounds that give the compounds certain characteristics.

homologous series A series of organic compounds that differ in length by one carbon atom.

hydrocarbon An organic compound containing only carbon and hydrogen.

hydroxyl group An –OH group characteristic of the organic family of alcohols.

isomers Two molecules with the same chemical formula but different structures.

ketone A family of organic compounds that contain the C=O functional group sandwiched between two other carbon atoms.

linear structure A chemical structure in which the atoms that compose the structure fall on a continuous line.

macromolecule A very large molecule, such as a polymer, typically containing thousands of atoms.

molar mass The mass of one mole of an element or compound.

monomer Repeating unit in a polymer.

organic compounds Compounds that contain carbon bonded to hydrogen, nitrogen, oxygen, or sulfur.

paraffin wax A soft, flammable solid composed of a mix of long chain hydrocarbons.

progesterone A female sex hormone with the formula $C_{21}H_{30}O_2$ that plays a role in the menstrual cycle and pregnancy.

testosterone A male sex hormone with the formula $C_{19}H_{28}O_2$.

ORGANIC CHEMISTRY & VITALISM

3-SECOND NUCLEUS

Vitalism, the belief that living things contained some unique force not contained in nonliving things, was proved false and abandoned in the nineteenth century.

3-MINUTE VALENCE

The downfall of vitalism is significant because it meant that living things could be studied from a chemical point of view. The reactions that happen in living organisms are not qualitatively different from those that happen outside of living organisms. The revolution that has occurred in biology over the last half-century is largely due to understanding life from a molecular perspective.

Early chemists divided compounds into two distinct types: organic and inorganic. Organic compounds were isolated from living organisms and tended to be quite fragile. For example, sugar is an organic compound (isolated from sugar cane or the sugar beet). If you heat sugar in a pan, it quickly decomposes. Inorganic compounds came from the Earth and tended to be more durable. For example, table salt is an inorganic compound (isolated from salt mines or from oceans). If you heat salt in a pan, the salt does not decompose—you simply get hot salt. Furthering the divide, early chemists were able to synthesize inorganic compounds in the laboratory, but not organic compounds. This divide fitted well with an eighteenth-century belief called vitalism, which suggested that all living organisms contained a vital force that separated them from nonliving things. The vital force allowed living organisms to synthesize organic compounds, but a chemist could not synthesize an organic compound in a beaker because no vital force was present. Vitalism died a slow death in the middle of the nineteenth century because chemists began to synthesize organic compounds from inorganic ones. Today, hundreds of thousands of organic compounds have been synthesized, including many compounds central to life itself.

RELATED TOPICS

See also
HYDROCARBONS
page 106

ALCOHOLS
page 108

ALDEHYDES, KETONES & ESTERS
page 110

3-SECOND BIOGRAPHIES

GEORG ERNST STAHL
1659–1734
German chemist and advocate of vitalism

FREDERICH WÖHLER
1800–82
German chemist who, in 1828, synthesized urea, an organic compound, from inorganic starting materials

EXPERT

Nivaldo Tro

Urea (center) was one of the first organic compounds synthesized from inorganic compounds.

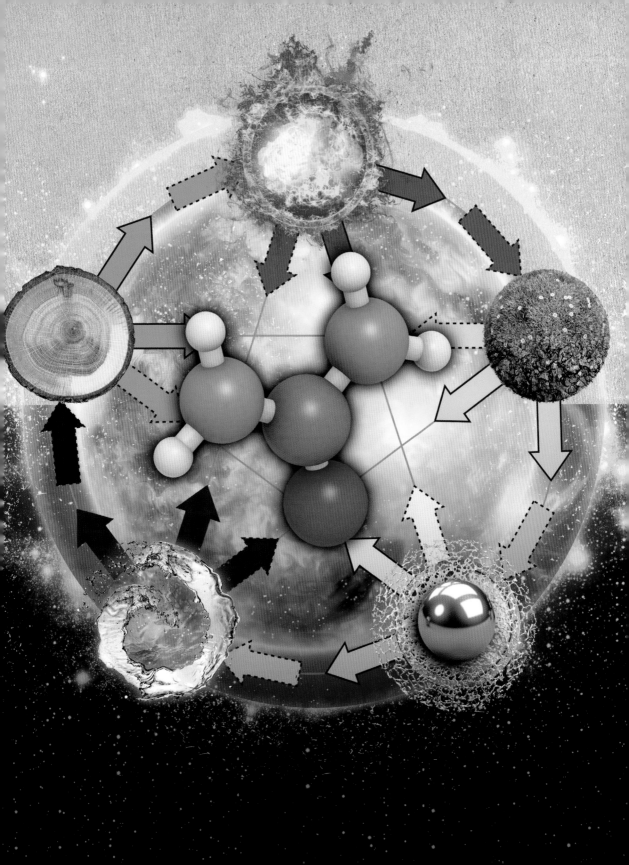

HYDROCARBONS

Hydrocarbons, the principle constituents of petroleum and natural gas, are organic compounds of hydrogen and carbon. Because carbon can bond, not only to other atoms like hydrogen, but also to itself (a process called "catenation"), there are many naturally occurring and synthetically produced hydrocarbon compounds. Hydrocarbons have had an unmatched influence on the economic, social, and environmental development of modern society. They are a convenient source of energy and the starting materials to make many common products such as polymers, textiles, and pharmaceuticals. Hydrocarbons in all physical forms find their way into our lives every day—in gas (natural gas), liquid (gasoline), and solid (paraffin wax) forms. They are, in general, classified as "saturated" (only single covalent bonds are present), "unsaturated" (at least one double or triple covalent bond exists between carbon atoms), or "aromatic" (at least one aromatic ring is present). Long chains of hydrocarbons can form linear, cross-linked, or cyclic structures, with the number of carbon atoms ranging from one (methane) to thousands. Hydrocarbons have high potential energy, which can be released upon combustion—generating energy convenient for applications such as transportation and heating.

RELATED TOPICS
See also
CARBON: IT'S NOT JUST
FOR PENCILS
page 96

ORGANIC CHEMISTRY
& VITALISM
page 104

3-SECOND BIOGRAPHIES
JAMES YOUNG
1811–83
Scottish chemist known as the father of the petrochemical industry who first distilled paraffin, a saturated hydrocarbon, from petroleum

EDWIN L. DRAKE
1819–80
An American railroad conductor credited with drilling the first modern oil well, near Titusville, Pennsylvania

EXPERT
Ali O. Sezer

Petroleum is separated into its hydrocarbon components due to the differences in the boiling points of those components.

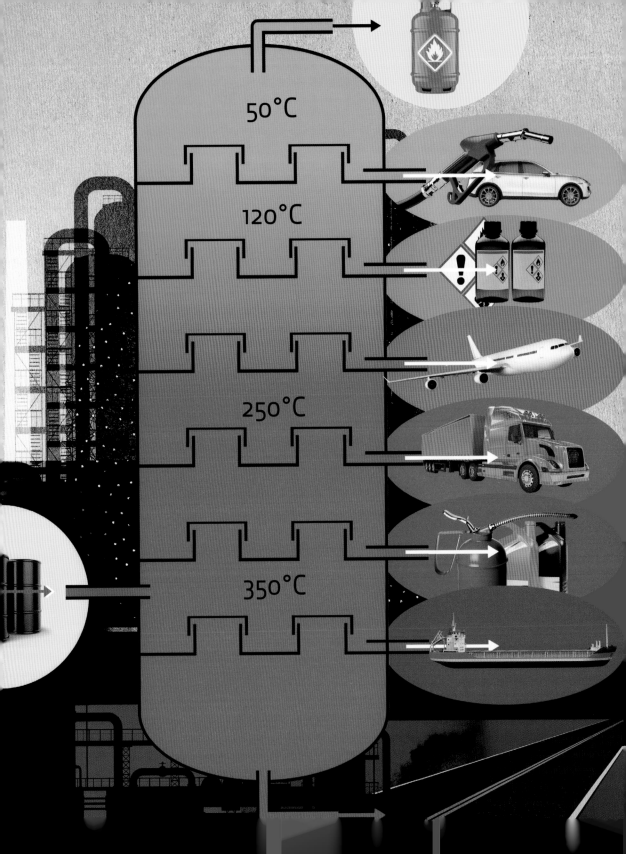

50°C

120°C

250°C

350°C

ALCOHOLS

Isopropanol (C_3H_7OH), a popular
antiseptic known as rubbing alcohol, and
ethanol (C_2H_5OH), the intoxicating component
of alcoholic drinks, belong to a very important
class of organic compounds called alcohols.
Alcohols contain the hydroxyl (–OH) functional
group. The smaller alcohols are clear, volatile,
and flammable liquids with a biting odor. The
alcohol family contains a homologous series
of compounds from hydrocarbons in which a
hydrogen atom is replaced by an –OH group.
Methanol, ethanol, and propanol are the
first three members of this series. The –OH
functional group is directly bonded to a carbon
atom, which is also connected to one (primary
alcohol), two (secondary alcohol), or three
(tertiary alcohol) other carbon atoms. Ethanol
and isopropanol are examples of primary and
secondary alcohols, respectively. The –OH group
makes alcohols highly polar, and the smaller
ones are all miscible with water—meaning that
they form homogenous solutions with water
in all proportions. Water-solubility decreases
noticeably as the number of carbon atoms
(molecular mass) increases. Today, alcohols find
a wide range of uses, from perfume-making
through food and pharmaceutical products
to medical applications.

EXPERT
Ali O. Sezer

*Alcohols are a common
family of organic
compounds that
includes substances
such as ethanol (the
alcohol in alcoholic
beverages).*

ALDEHYDES, KETONES & ESTERS

The pleasant odor and flavor

of fresh almonds and peppermint are mainly due to benzaldehyde and menthone, two naturally occurring compounds whose molecule contains the carbonyl functional group (–C=O). Benzaldehyde is an example of a family of organic compounds called aldehydes. In an aldehyde, the –C=O group is at the end of a carbon chain. Menthone is an example of a family of organic compounds called ketones. In a ketone, the carbonyl group is in the middle of the carbon chain. Another common family of organic compounds is the ester family. In an ester, an oxygen atom interrupts the bonding of the carbonyl group to the carbon chain (–CO$_2$C–). Benzyl acetate, a component in the smell of strawberries, pears, and jasmine, is an ester. Small aldehydes and ketones (those with low molar masses), such as formaldehyde (an important industrial solvent) and acetone (nail-polish remover), have a strongly pungent odor. But aldehydes with increasing molar masses generally have more pleasant and fruity odors. Aldehydes, ketones, and esters are commonly found naturally and manufactured industrially as odors and flavors in food and pharmaceutical products. Many solvents used in adhesives, paints, perfumes, plastics, and fabrics also contain aldehydes, ketones, and esters.

3-SECOND BIOGRAPHIES
LEOPOLD GMELIN
1788–1853
German chemist who first introduced the terms "ester" and "ketone"

JUSTUS VON LIEBIG
1803–73
German chemist considered the father of organic chemistry, who first used the term "aldehyde"

EXPERT
Ali O. Sezer

Many of the odors you smell every day come from the naturally occurring families of aldehydes, ketones, and esters.

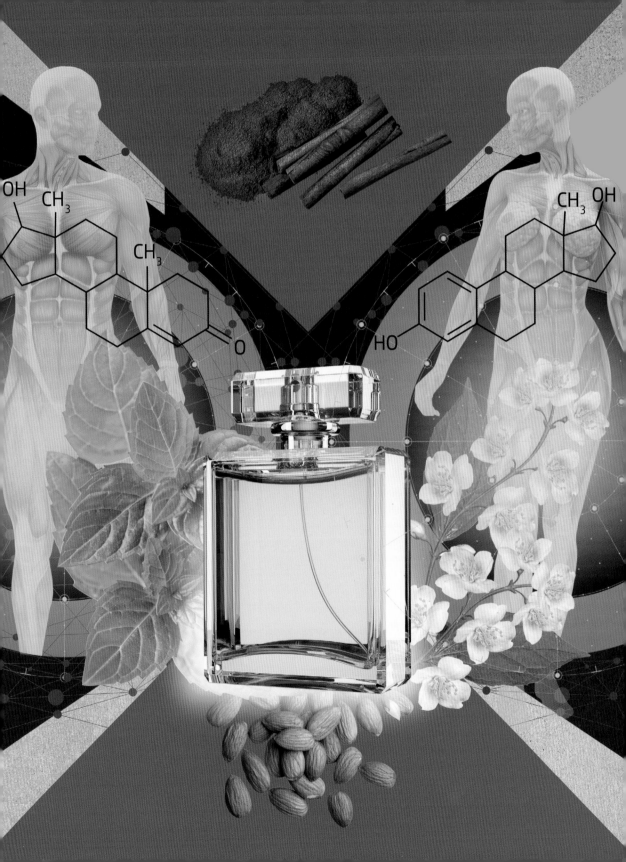

CARBOXYLIC ACIDS & AMINES

3-SECOND NUCLEUS
Carboxylic acids and amines are naturally occurring acids and bases, respectively.

3-MINUTE VALENCE
The reaction between a carboxylic acid and an amine is an acid-base reaction that links the two molecules together and forms water as a by-product. This reaction is important in biochemistry because it is responsible for the linking of amino acids—which are molecules that have a carboxylic acid group on one side and an amine group on the other —to form proteins, the workhorse molecules in living organisms.

Carboxylic acids and amines are important and familiar families of organic compounds. Carboxylic acids are organic acids and can be identified by their –COOH functional group. Like all acids, carboxylic acids taste sour. Acetic acid, the active component of vinegar, is a carboxylic acid. The word vinegar originates from the French for wine (*vin*) and sour (*aigre*). When wine is left exposed to air, the ethanol oxidizes to form carboxylic acid, ruining the wine. Citric acid is another carboxylic acid; it is responsible for the sour taste of lemons and limes. Amines are organic bases and contain a nitrogen atom bonded to one or more carbon atoms. Many amines have an unpleasant odor. When living things die, their proteins are broken down into amines, which evaporate into the air. For example, the smell of rotting fish is due to trimethylamine, and that of decaying animal flesh to caderverine, both foul-smelling amines. Some plant-based amines, known as alkaloids, have the ability to alter sensory pathways. Caffeine, nicotine, and cocaine are all alkaloids that stimulate the central nervous system, resulting in feelings of increased alertness and energy. Other alkaloids, such as morphine and codeine, have the opposite effect: they depress the central nervous system. Morphine is a powerful depressant used to treat extreme pain.

RELATED TOPICS
See also
ORGANIC CHEMISTRY & VITALISM
page 104

ALCOHOLS
page 108

ALDEHYDES, KETONES & ESTERS
page 110

3-SECOND BIOGRAPHY
HERMANN KOLBE
1818–84
German chemist who greatly contributed to the development of organic chemistry and was the first to synthesize acetic acid

EXPERT
Nivaldo Tro

Acetic acid (top) is responsible for the smell of vinegar. Trimethylamine (bottom) is responsible for the foul smell of rotten fish.

31 July 1800
Born in Eschersheim, Germany

1823
Receives a medical degree from the University of Heidelberg, Germany

1827
Prepares the first pure sample of aluminum from its compounds

1828
Discovers how to make urea synthetically in the laboratory and isolates the elements beryllium and yttrium

1832
Starts working as a chemistry professor at the University of Göttingen, Germany

1834
Elected foreign member of the Royal Swedish Academy of Sciences

1854
Elected foreign member of the Royal Society of London

1855
Elected member of the Royal Academy of Berlin

1862
Produces calcium carbide and acetylene gas

1872
Awarded the Copley Medal by the Royal Society of London

23 September 1882
Dies in Göttingen, Germany

FRIEDRICH WÖHLER

Friedrich Wöhler is known not only for his pioneering research in chemistry but also for his model teaching laboratories that revolutionized how experimental chemistry is taught around the world today. His interest in chemistry and mineralogy developed in the early years of his education in Frankfurt, Germany. Although Wöhler went on to receive a medical degree from the University of Heidelberg, his real interest was always in chemistry. Leopold Gmelin, one of the best-known chemists of the nineteenth century, was Wöhler's instructor at Heidelberg. He quickly realized that Wöhler was too advanced for his courses and sent him to study with Jons Jacob Berzelius, the world-famous Swedish chemist who is considered to be one of the fathers of modern chemistry.

Wöhler spent a year studying mineralogy with Berzelius. He not only received the best available education but also developed a lifelong friendship with his tutor. Wöhler later translated much of Berzelius's work into German, including his well-known *Textbook of Chemistry*. Wöhler himself wrote a number of textbooks on organic and inorganic chemistry later in his career, including *Outlines of Organic Chemistry* in 1840.

Wöhler returned to Germany in 1825, and took a position at the newly founded Berlin Gewerbeschule (trade school), where he conducted groundbreaking research that won him international fame. In 1827 he isolated pure aluminum for the first time from its compounds. The following year, he announced his second discovery in a famous letter to Berzelius, where he explained how he had synthetically isolated urea in the laboratory without the need of a living kidney, and that this compound had the same chemical composition as another compound called ammonium cyanate. This discovery was significant, as researchers then believed that a "vital force" in living things was necessary to make organic compounds.

Two years after Wöhler's letter, Berzelius explained these discoveries and coined the term isomerism, a tremendously important concept in modern chemistry. From 1832 until his death in 1882, Wöhler served as a chemistry professor at the University of Göttingen, where he developed, in parallel with Justus Freiherr von Liebig at the University of Giessen, today's widely adopted laboratory-based methods of science teaching. Wöhler is also credited with having inaugurated the tradition of establishing scientific research groups in which students can also carry out research activities.

Ali O. Sezer

CHEMISTRY
COPYING NATURE

The Pacific yew tree is rather unremarkable. It grows in the Pacific Northwest to heights of about 30–50 ft (10–15 m) and has flat, green needles and red berries. However, its bark contains the miracle drug Taxol, now used to treat ovarian, lung, breast, and colon cancer. The biological action of the yew tree has been known since Greek times, and Native Americans used the tree for medicinal purposes. Because of this known history, researchers in the 1960s included the tree in large-scale screening for cancer-fighting agents. The positive results eventually led researchers to isolate the active ingredient Taxol. But the amount of Taxol needed to treat a single cancer patient required the harvesting of several trees that were around one hundred years old, creating environmental problems. As often happens with natural products, however, researchers came up with another way to obtain the compound: taxol is now synthesized from a precursor found in the needles of the European yew tree. Because the trees don't have to be cut down to harvest the needles, this route is sustainable. Today, millions of cancer patients have benefited from this natural product. The story of Taxol is representative of the field of natural products research, an active area that produces many novel and useful compounds.

3-SECOND NUCLEUS
Plants, animals, and microbes are sources of valuable and often healing chemical compounds.

3-MINUTE VALENCE
Natural products researchers scour living organisms for the useful chemicals they might contain. For example, penicillin (an antibiotic) was extracted from microbes, and the early form of aspirin was extracted from the bark of the willow tree. Often, chemists later find ways to synthesize these compounds in the laboratory, but the myriad of compounds in nature often shows us the way.

RELATED TOPICS
See also
ORGANIC CHEMISTRY
& VITALISM
page 104

BIOTECH DRUG SYNTHESIS
page 134

3-SECOND BIOGRAPHIES
MONROE ELIOT WALL
1916–2002
American chemist credited with the co-discovery of Taxol

MANSUKH C. WANI
1925–
Indian American chemist credited with the co-discovery of Taxol

EXPERT
Nivaldo Tro

The yew tree is the source of the cancer-fighting agent Taxol.

POLYMERS

Polymers permeate most of

human life. It is hard to imagine society without them, given the significant role natural and synthetic polymers play—from medicine to food packaging, and from clothing to housewares. German chemist Hermann Staudinger first demonstrated the existence of macromolecules (polymers) as large, chainlike molecules of smaller repeating units called monomers. Humans have known natural polymers, such as cotton and rubber, for thousands of years. But their chemical structure was long subject to debate until Staudinger proved the macromolecular structure of natural rubber— a Nobel Prize–winning discovery in 1953. The term polymer is derived from the Greek *poly meros*, meaning "many parts." Many molecules can act as monomers, making it possible to create a wide range of polymeric materials with desired characteristics. Polyethylene, for example—one of the most common plastic materials found in packaging bags and bottles —is a chainlike molecule consisting of an ethylene monomer backbone. Polymers can have high molecular mass as monomers can link in a variety of ways to grow into very large molecules. Some are light, hard, strong, and flexible; others exhibit unique chemical, thermal, electrical, and optical characteristics.

3-SECOND NUCLEUS

Polymers—chainlike macromolecules consisting of chemically linked repeating units (monomers)—play an unmatched role providing convenience in many aspects of human life.

3-MINUTE VALENCE

Polymers are in general known as electrical insulators. However, some organic polymers— consisting of monomer units linked by alternating single and double bonds in the carbon backbone— exhibit a low, inherent electrical conductivity, which can be significantly improved by chemically mixing in electron-donating and/or -receiving agents, a process called doping. Doped polymers, such as organic light emitting diodes (OLED), have revolutionized the electronics industry.

RELATED TOPICS

See also
THE LEWIS MODEL FOR CHEMICAL BONDING
page 32

HYDROCARBONS
page 106

3-SECOND BIOGRAPHIES
HERMANN STAUDINGER
1881–1965
German chemist who won the 1953 Nobel Prize in Chemistry for discovering the macromolecular structure of natural rubber

HIDEKI SHIRAKAWA
1936–
Japanese chemist who won the 2000 Nobel Prize in Chemistry for codiscovering the existence of conductive polymers

EXPERT
Ali O. Sezer

Polymers are chainlike molecules that compose a range of substances such as plastics.

BIOCHEMISTRY

BIOCHEMISTRY
GLOSSARY

acid group A functional group consisting of –COOH.

adenosine triphosphate A biomolecule with the formula $C_{10}H_{16}N_5O_{13}P_3$ that is used as a main vessel for energy transport in living organisms.

alkanes Hydrocarbons with the general formula C_nH_{2n+2}.

amine group In an amino acid, the $-NH_2$ group.

amino acids The individual units that, when linked together in a specified order, form proteins. Amino acids have a central carbon atom, an amine group, an acid group, and a side chain (that varies in structure from one amino acid to another).

base pair The two parts of nucleic acids that uniquely pair together to form the double helix in DNA and allow precise copying. In DNA, adenine pairs with thymine and guanine pairs with cytosine.

carboxylic acid group The –COOH group in organic and biochemical compounds. The group is polar and acidic.

cellulose A complex carbohydrate composed of repeating glucose units. Cellulose is the main structural material in plants.

DNA Deoxyribonucleic acid, a biomolecule composed of repeating units (called nucleotides) responsible for carrying the genetic information in all known living organisms.

disaccharides A class of sugars composed of two monosaccharides linked together.

esters A class of organic compounds consisting of a –COO– group sandwiched between two or more carbon atoms.

flash photolysis A technique to study light-activated chemical reactions in which a flash of light is used to initiate the chemical event, which is then monitored as a function of time.

genome The complete set of genetic material of an organism.

glucose A carbohydrate with the formula $C_6H_{12}O_6$ that circulates in the blood of animals and humans.

hormones Biochemical compounds that are transported in the blood to targets where they stimulate and regulate biochemical processes.

modular polymers Polymers that can be built up one unit (or monomer) at a time.

monosaccharides A carbohydrate composed of three to eight carbon atoms and one aldehyde or ketone group.

nonpolar A substance composed of molecules with uniform charge distribution. Nonpolar substances generally do not mix well with water.

nucleotide The individual unit that, when linked with other nucleotides, forms a nucleic acid (such as DNA). Each nucleotide contains a phosphate group, a sugar, and a base.

polar A substance composed of molecules with an asymmetric charge distribution.

polynucleotide A chain of nucleotides bonded together found in hereditary molecules such as DNA and RNA.

recombinant DNA Synthetic DNA that contains genetic material from different sources.

ribonucleotide The monomer that, when linked to other ribonucleotides, forms RNA.

sucrose A carbohydrate with the formula $C_{12}H_{22}O_{11}$.

triglyceride A type of fat that has a three-carbon backbone with three fatty acids attached (one to each carbon atom).

CARBOHYDRATES

Carbohydrates are so named

because their general formula is a multiple of one carbon atom and one water molecule, $(CH_2O)_n$. Structurally, the carbon atoms are arranged in a ring (that can interconvert into a straight chain) and have multiple hydroxyl (OH) groups attached, making simple carbohydrates polar and therefore soluble in water. The ability to dissolve in water is important to one of the main functions of carbohydrates: storing and transporting energy for living organisms. The carbohydrate glucose ($C_6H_{12}O_6$) is typical. It must be easily transported in the blood to places in the body where energy is being used. Carbohydrates such as glucose (which are also called monosaccharides, meaning one sugar) can link together to form disaccharides, such as sucrose ($C_{12}H_{22}O_{11}$), which is table sugar. They can also link together to form long, chainlike molecules called complex carbohydrates such as starch, glycogen, and cellulose. Starch (think potatoes) is the main energy storage medium for plants, and glycogen is used by animals as a compact way to store glucose in the muscles. Cellulose is the most common organic substance on Earth. It is more rigid than the other complex carbohydrates and is the main structural component in plants.

RELATED TOPICS
See also
LIPIDS
page 126

AMINO ACIDS & PROTEINS
page 128

3-SECOND BIOGRAPHIES
ANDREAS MARGGRAF
1709–82
German chemist who first isolated glucose from raisins

EMIL HERMANN FISCHER
1852–1919
German chemist and winner of the 1902 Nobel Prize in Chemistry for his pioneering work on sugars

EXPERT
Nivaldo Tro

3-SECOND NUCLEUS
Carbohydrates are multicarbon aldehydes or ketones with many OH groups attached; they act as short-term energy stores and the main structural components of plants.

3-MINUTE VALENCE
Carbohydrates are common in the foods we eat. Monosaccharides can pass directly through our intestinal wall and enter the bloodstream as ready sources of energy. Disaccharides and complex carbohydrates, however, must be broken down into monosaccharides before they can pass into the bloodstream. Our bodies can break down sugars and starches, but we lack the enzyme to break down cellulose (also known as dietary fiber), which is why cellulose passes through the digestive tract, giving bulk to stools and preventing constipation.

Carbohydrates include simple sugars such as glucose (top) and complex carbohydrates such as cellulose (bottom).

LIPIDS

Lipids are the only biomolecule

defined by what they are not: lipids are not able to dissolve in water. This insolubility enables lipids to form thin, oily membranes and to clump together into oily droplets that serve as high-density stores of metabolic energy. In fact, many lipids contain a large, burnable hydrocarbon group similar to the alkanes in gasoline. In fatty acids (a type of lipid), for instance, a long hydrocarbon chain is attached to a single polar carboxylic acid group. In tryglycerides (another type of lipid) three long hydrocarbon chains are attached to a short three-carbon head. This highly nonpolar structure causes tryglycerides to glob together into oily "fat" droplets. Another type of lipid has only two long hydrocarbon chains attached to a more polar head (a three-carbon unit containing a phosphate group). The result is a rodlike molecule with a charged water-loving "head" and oily tail. These form sheets with the tails lined up in oily sheets on the one side and the "heads" all facing to the other. To keep the oily surface of the tails out of water, two sheets line up to give a bilayer membrane, with the tails on the inside and the heads forming the water-facing surfaces. These bilayer membranes are the fundamental barriers that encapsulate living cells.

3-SECOND NUCLEUS

Lipids' insolubility in water enables them to form extended membranes that enclose biological solutions and to function as particularly dense stores of metabolic energy.

3-MINUTE VALENCE

Even though they cannot be built up into modular polymers in the same way that other biomolecules can, lipids perform many varied biological functions. Lipid membranes act as barriers between the insides and outsides of cells while triglyceride "fats" function as long-term energy stores in plants and animals. Other lipids called hormones act as biological messengers, being secreted by glands and carried to target cells where they trigger a physiological response.

RELATED TOPICS

See also
THE FORCES THAT HOLD MATTER TOGETHER
page 42

HYDROCARBONS
page 106

CARBOXYLIC ACIDS & AMINES
page 112

3-SECOND BIOGRAPHIES

MICHEL CHEVREUL
1786–1889
French chemist who was a pioneer in studying the chemistry of soaps, fats, and oils

CHARLES ERNEST OVERTON
1865–1933
British biologist who first proposed that lipids might act as a cell membrane

EXPERT

Stephen Contakes

One of the many functions of lipids is to encapsulate cells by forming a bilayer.

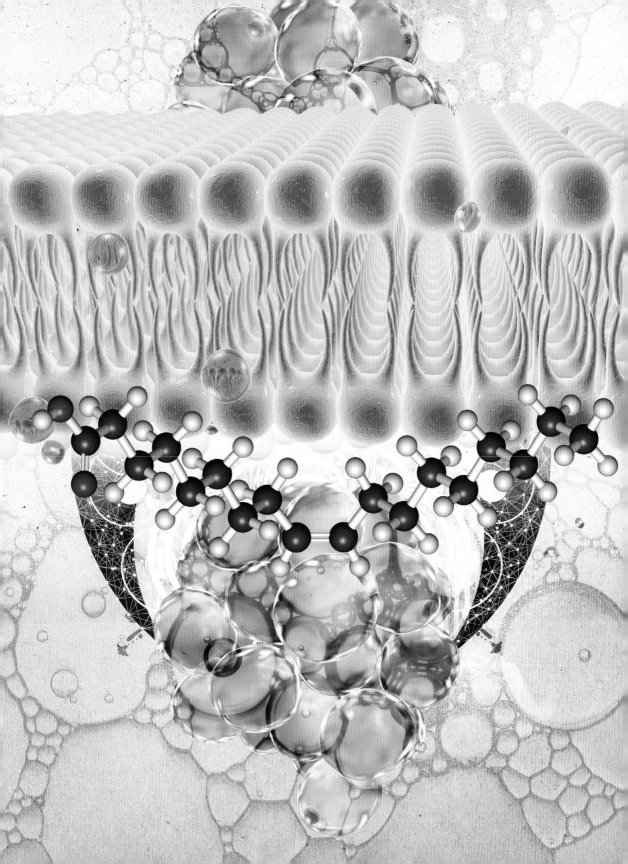

AMINO ACIDS & PROTEINS

3-SECOND NUCLEUS
Amino acids can be
strung together into
protein chains that fold
into a diverse array of
shapes and carry out many
biological functions.

3-MINUTE VALENCE
When you take a breath,
the oxygen you breathe
becomes bound to a
protein called hemoglobin
in your red blood cells,
which transports the
oxygen to your muscles
and other tissues where
it is used to "burn" fats
and carbohydrates in
interconnected sequences
of protein-catalyzed
reactions. These generate
the energy your body
needs to move, synthesize
other biomolecules, and
produce the electrical
signals needed for nerve
cells to function.

Proteins are chainlike

biomolecules that carry out a bewildering array of functions. Some, like the collagen in your skin, serve as structural supports. Others, like the motor protein myosin, enable muscles to relax and contract. Others adopt compact "globular" shapes and can store or transport smaller molecules around, control cellular metabolism by speeding up particular chemical reactions, or even recognize and bind other molecules. Some proteins, like insulin, act as intercellular signals, while others regulate bodily processes by chemically modifying other proteins to adjust how they function. Yet proteins are comprised of only 20 basic building blocks called amino acids, so named because they all contain a central carbon atom bonded to an amine group and an acid group. The central carbon is also bonded to a third variable group called a side chain, which may be polar, nonpolar, small, large, acidic, or basic. Amino acids can string together to form long polymers whose properties can vary widely based on the exact sequence of these side chains. Interactions between the side chains at different points along the polymer with each other and with surrounding water causes the protein to wriggle and fold into specific shapes, which in turn determine the functions they can perform.

RELATED TOPICS
See also
REACTION RATES &
CHEMICAL KINETICS
page 70

CARBOXYLIC ACIDS & AMINES
page 112

CHEMISTRY COPYING NATURE
page 116

3-SECOND BIOGRAPHIES
GERARDUS JOHANNES
MULDER
1802–80
Dutch chemist who first
described the composition
of proteins

JOHN KENDREW
& MAX PERUTZ
1917–97 & 1914–2002
British biochemist and
Austrian-born molecular
biologist who determined the
first 3-D structures of proteins

EXPERT
Stephen Contakes

*Some proteins fold
into globular shapes
(center), while others
have more linear
structures (bottom).*

23 December 1912
Born in Benton City, Missouri

1933
Receives a BA degree in chemistry from the University of Missouri-Columbia

1935
Receives a BA degree in education from the same institution

1937
Receives an MA degree in chemistry, again from the University of Missouri-Columbia

1940
Receives a PhD degree in physical chemistry, once again from the University of Missouri-Columbia

1945
Joins the chemistry department at Mount Holyoke College in Massachusetts

1960
Receives the Citation of the University of Missouri College of Arts and Sciences

1969
Receives the Manufacturing Chemists Association Award in College Chemistry Teaching

1982
Receives the Chemical Education Award from the American Chemical Society

1978
Becomes the first female president of the American Chemical Society

1983
Serves as the president of the American Association for the Advancement of Science

1989
Coauthors a textbook with Edwin S. Weaver entitled *Chemistry: A Search to Understand*

8 August 1998
Dies in Holyoke, Massachusetts

ANNA J. HARRISON

Anna Jane Harrison was an American chemist and educator who believed in the importance of improving science education and increasing public awareness of science. Born to a family of farmers in Benton City, Missouri, Harrison took an interest in chemistry as early as elementary school and by high school this had turned to fascination. She received all her advanced degrees from the University of Missouri-Columbia, including two BA degrees (in chemistry and education) and a PhD in physical chemistry. After five years of teaching at the Sophie Newcomb College of Tulane University in New Orleans, she took a position as a chemistry professor at Mount Holyoke College in Massachusetts, where she taught until her retirement in 1979. She continued to teach even after her retirement, at the U.S. Naval Academy in Annapolis, Maryland.

At Mount Holyoke College, Harrison had a chance to work with renowned chemistry professor Emma Perry Carr on the spectroscopic study of molecular structure. She carried out research using a technique called flash photolysis to study chemical reactions by monitoring the dissociation and association of different molecular compounds. Her research activities also included work carried out with the A.J. Griner Company of Kansas City on field kits for the detection of toxic smoke for soldiers in the Second World War.

Harrison is perhaps better known today for her skills in science education. She is credited with helping change the chemistry profession from being a male-dominated profession to one that is more welcoming of diversity in gender, race, and ethnicity. More than one hundred years after its foundation, the American Chemical Society elected Harrison as its first female president in 1978. Harrison had a natural talent for making complicated concepts clear and comprehensible to her students. Her approach to education included helping students acquire better knowledge of how to make good public policy decisions related to science. In the 1970s, Harrison became an outspoken advocate of improved communication of science to the public, particularly to public officials. She served on many advisory boards including the National Science Board, and traveled to different parts of the world sharing her experience in public education of science.

Ali O. Sezer

THE BIOLOGICAL BLUEPRINT: NUCLEIC ACIDS

3-SECOND NUCLEUS

Nucleic acids form chains of alternating phosphates and carbohydrates with attached wedgelike nitrogen-containing bases, the sequence of which encodes biological information.

3-MINUTE VALENCE

Life's information-bearing molecule is deoxyribonucleic acid or DNA. DNA contains within it the chemical code for protein synthesis and is passed from parent to offspring, which is why you have characteristics similar to your parents. In 2003, scientists successfully mapped the entire human genome, a chemical code containing about three billion units (base pairs).

DNA is a long, chainlike molecule containing units called nucleotides. Each nucleotide unit contains a negatively charged phosphate group attached to a carbohydrate ring, which is itself attached to a wedgelike nitrogen-containing group called a base. Bases come in four varieties, all of which are flat and nonpolar on top and bottom, but have specific patterns of polar nitrogen, oxygen, and hydrogen atoms along their edges. These patterns allow bases to recognize "complementary" bases, namely those that have the right pattern of polar groups to interact strongly, giving a base pair. Therefore, when deoxyribonucleic acids' sugar-phosphate groups are linked together into long polynucleotide chains, the bases along the chain can generate another polynucleotide strand with a sequence of bases complementary to the first. Some base sequences in DNA encode instructions for making proteins. These, along with nearby base sequences that tell the cell's machinery when to make those proteins, make up the units of heredity called genes. However, nucleic acids aren't only used to store and transmit genetic information. The cell's main energy currency, adenosine triphosphate (ATP), is a ribonucleotide in which the phosphate has been replaced by a chain of three linked phosphate groups.

RELATED TOPICS

See also
THE FORCES THAT HOLD MATTER TOGETHER
page 42

CARBOXYLIC ACIDS & AMINES
page 112

CARBOHYDRATES
page 124

3-SECOND BIOGRAPHIES
OSWALD AVERY
1877–1955
Canadian-born medical researcher who demonstrated that DNA is genetic material

JAMES WATSON & FRANCIS CRICK
1928– & 1916–2004
American and British molecular biologists who determined DNA's double-helical structure

EXPERT
Stephen Contakes

DNA has a double helical structure in which complementary bases connect along the middle.

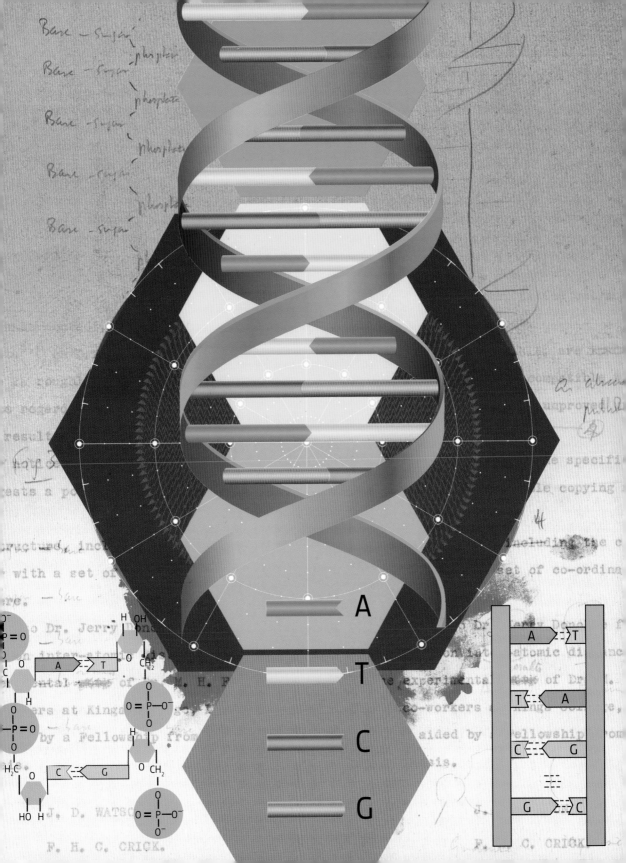

BIOTECH DRUG SYNTHESIS

3-SECOND NUCLEUS

Human proteins can be synthesized by inserting the human gene for the desired protein into bacterial, plant, or animal cells. As these cells grow and divide, they synthesize the desired protein.

3-MINUTE VALENCE

Genetic engineering—the process of modifying an organism's genome for a particular purpose—has been used, not only to produce lifesaving medicines but also to produce animals or plants with desirable characteristics. Genetic modification of soybeans, tomatoes, and rice has resulted in crops with more resilience and higher nutritional value. In spite of rigorous scientific testing, the controversy surrounding some of these products has resulted in increased scrutiny of their safety.

Before 1922, diabetes was fatal. Then a fourteen-year-old diabetes patient on the verge of death was given insulin (a protein that regulates blood sugar) derived from animal sources. The patient recovered—and survived. Soon insulin (harvested mostly from pigs) became available for widespread use, changing diabetes into a manageable long-term disease. However, some patients did not tolerate pig insulin very well. In the 1980s a company called Genentech figured out a way to synthesize human insulin by inserting the gene for human insulin into the DNA of bacterial cells. When the bacteria reproduced, they copied the human insulin gene and passed it on to their offspring. Furthermore, as the genetically modified bacteria synthesized the proteins they needed to survive and reproduce, they also synthesized human insulin. The insulin was harvested from the bacterial cultures, purified, and administered to diabetics. Today, diabetics take human insulin, synthesized in this way. The DNA instructions for making desired proteins can also be inserted into the DNA of plants or animals. For example, in 2015 the FDA approved a drug to treat Wolman disease, a rare but fatal disease caused by a deficiency of an enzyme called LAL. The drug is harvested from the eggs of chickens that have been genetically modified to produce LAL.

RELATED TOPICS

See also
AMINO ACIDS & PROTEINS
page 128

THE BIOLOGICAL BLUEPRINT:
NUCLEIC ACIDS
page 132

3-SECOND BIOGRAPHIES

FREDERICK BANTING
1891–1941
Canadian physician who was awarded the 1923 Nobel Prize in Medicine for his discovery of insulin

FREDERICK SANGER
1918–2013
British biochemist who was awarded the Nobel Prize in Chemistry in 1958 for his determination of the structure of insulin

PAUL BERG
1926–
American biochemist awarded the 1980 Nobel Prize in Chemistry for his work on recombinant DNA technology

EXPERT

Nivaldo Tro

Human insulin is synthesized by the genetic modification of bacteria.

NUCLEAR CHEMISTRY

alpha particle A type of particle given off in one type of radioactive decay. An alpha particle contains two protons and two neutrons and is symbolized as $^4_2$He.

amu A unit of mass used for subatomic particles. 1 amu = 1.66 x 10^{-27} kg.

beta particle A type of particle given off in one type of radioactive decay. A beta particle is an electron and is symbolized as $^{\;\;0}_{-1}e$.

chemical reaction A process in which the atoms in one or more substances (the reactants) rearrange to form different substances (the products).

critical mass In nuclear fission, the minimum mass of fissile material needed to maintain a self-sustaining nuclear reaction.

Einstein's equation $E=mc^2$ An equation that relates mass to energy, which means that the two are interconvertible.

electron A subatomic particle with a negative charge and a mass of 0.00055 amu.

gamma ray A high-energy photon given off in one type of radioactive decay and often in conjunction with other types. A gamma ray is symbolized as $^0_0\gamma$.

gene A strand of DNA that codes for a single protein.

isotope An atom that has the same number of protons as another atom, but a different number of neutrons.

metabolism The process by which living organisms convert certain compounds into the energy needed to survive and reproduce.

neutron A subatomic particle with no charge and a mass of 1 amu.

nuclear fission A nuclear reaction in which a large nucleus splits into smaller fragments and releases energy.

plutonium A synthetic chemical element with atomic number 94 used in nuclear chemistry, especially nuclear power and atomic bombs.

proton A subatomic particle with a positive charge and a mass of 1 amu.

radioactivity The emission of small energetic particles from the nuclei of certain unstable isotopes.

radiocarbon dating A method of determining the age of previously living material by measuring the C-14 content in the material.

radiopharmaceutical Pharmaceutical agents that are radioactive and used in the diagnosis and treatment of disease.

tracers A chemical compound where one or more atoms have been replaced with a radioactive isotope that allows a scientist to trace where the atom goes in a particular process.

uranium A radioactive chemical element with atomic number 92 used in nuclear chemistry, especially nuclear power and atomic bombs.

X-rays A form of electromagnetic radiation with wavelengths slightly longer than gamma rays and used to image bones and organs.

RADIOACTIVITY

3-SECOND NUCLEUS

Some atoms spontaneously "decay," producing alpha and beta particles and gamma rays that can penetrate through various materials. This is called radioactivity.

3-MINUTE VALENCE

Radioactive elements can be used as "tracers" to follow the pathway of a chemical reaction or monitor concentrations of elements in research, environmental, agricultural, and medical settings. Radioactive elements are also used to establish the ages of various objects including once-living systems (carbon-14), early humanoids (potassium-40), and the Moon, Earth, and various rocks and minerals (uranium and thorium isotopes).

As a scientific poet once wrote, "Atoms ... fly to bits with utmost facility." Counterintuitively, some atoms spontaneously fall apart to produce rays and particles that can penetrate through various materials, including metals and our own bodies. Antoine Becquerel and Pierre and Marie Curie, working in Paris in the last decade of the nineteenth century, were the first to recognize this phenomenon in uranium minerals. Madame Curie called this "radioactivity" (from the Latin *radius*, meaning "ray") and soon found two mysterious, previously unknown elements ("radium" and "polonium") that emitted more intense "radiation" than uranium itself. Initially, radioactivity was thought to exist in two types, designated logically enough by Ernest Rutherford as alpha and beta. (Gamma rays were discovered several years later.) Positively charged alpha particles, soon found to be energetic helium nuclei, He^{2+}, had less ability to penetrate various substances. Negatively charged beta particles, soon found to be energetic electrons, were much lighter and able to penetrate a variety of materials well. Gamma rays, the most penetrating of all, were high-energy electromagnetic radiation. Most remarkably, it is now clear that certain types of atoms spontaneously "decay" and spit out tiny charged particles and highly intense radiation.

RELATED TOPICS

See also
MATTER IS MADE
OF PARTICLES
page 14

THE STRUCTURE OF
THE ATOM
page 16

SPLITTING THE ATOM
page 142

3-SECOND BIOGRAPHY

WILLIAM RAMSAY
1852–1916
Scottish chemist who wrote the 1902 poem "The Death Knell of an Atom," which contains the following stanza:

"So the atoms, in turn, we now clearly discern,/Fly to bits with the utmost facility;/
They wend on their way, and in splitting, display/An absolute lack of stability."

EXPERT

Glen E. Rodgers

When atoms emit radiation they change their identity.

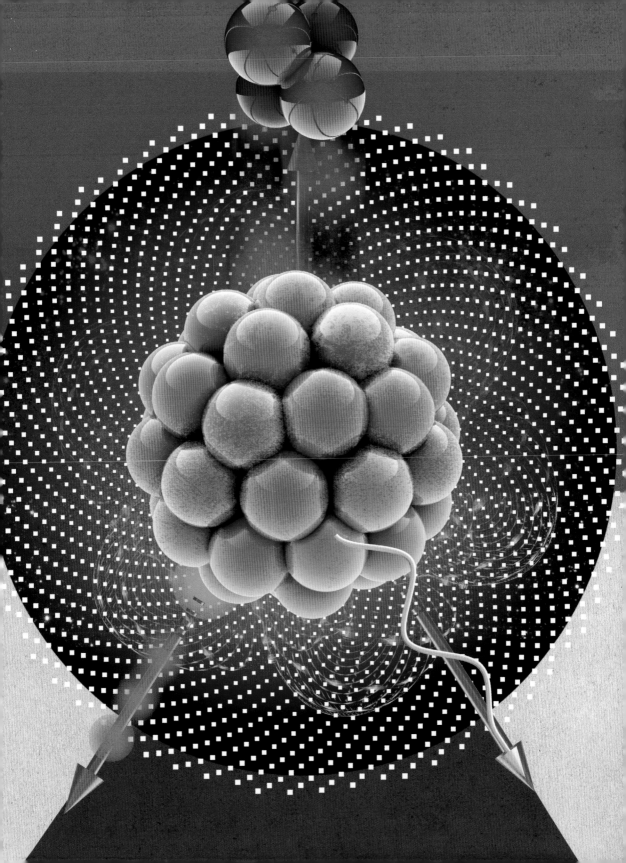

SPLITTING
THE ATOM

3-SECOND NUCLEUS

Firing neutrons into critical masses of fissionable materials splits the atoms apart and produces additional neutrons; the resulting chain reaction releases large amounts of energy.

3-MINUTE VALENCE

Why don't atomic nuclei, particularly those containing dozens of positively charged and therefore mutually repulsive protons, just burst apart? It turns out that some do. Large nuclei of certain types of atoms of uranium or plutonium atoms act like wobbly drops of liquid tenuously held together by a nuclear surface tension that can be easily disrupted.

In 1938, Otto Hahn shot neutrons at uranium atoms and was amazed to discover that they appeared to split roughly in half. He had discovered nuclear fission, whereby a large nucleus splits to form two smaller nuclei and several more neutrons. If a "critical mass" of a sufficiently pure isotope of uranium or plutonium is present, these additional neutrons can go on to hit other fissionable nuclei and cause a chain reaction which, as calculated using Einstein's equation $E=mc^2$, releases inordinate amounts of energy, far in excess of that obtained from conventional chemical reactions. This energy can be harnessed to create electricity (nuclear energy) or to create explosions (nuclear bombs). The uranium-based bomb dropped on Hiroshima, Japan, on August 6, 1945, was called "Little Boy." In this "gun-type" assembly, the critical mass was obtained by firing a uranium "bullet" into a hollow cylinder of uranium. The plutonium-based bomb dropped on Nagasaki, Japan, on August 9, 1945, was called "Fat Man." In this implosion-type assembly, the critical mass was obtained using a lens assembly that fired small bits of plutonium all toward the center of the bomb.

RELATED TOPICS

See also
THE STRUCTURE OF
THE ATOM
page 16

INSIDE THE ATOM
page 18

OTTO HAHN
page 148

3-SECOND BIOGRAPHIES

MARIE CURIE
1867–1934
Polish-born French chemist who developed the theory of radioactivity

OTTO HAHN
1879–1968
German winner of the 1944 Nobel Prize in Chemistry, who discovered nuclear fission

EXPERT
Glen E. Rodgers

In nuclear fission, a neutron causes an unstable nucleus to split, releasing large amounts of energy.

NUCLEAR WEIGHT LOSS

3-SECOND NUCLEUS

Protons and neutrons each lose a tiny bit of mass when nuclear fission occurs. This mass is converted into energy.

3-MINUTE VALENCE

If matter and energy are two sides of the same coin, then is it appropriate to imagine protons and neutrons as hard spheres? Is it better to think of them as little balls of energy? Because matter and energy are interconvertible, perhaps it doesn't really matter. This is another indication that the subatomic universe is a very strange place.

When an atom splits, either in a nuclear power plant or during the detonation of a nuclear weapon, a tremendous amount of energy is released. Where does all of this energy come from? The answer lies in Einstein's remarkably simple equation that relates energy and matter: $E=mc^2$—energy equals mass times the speed of light squared. This equation states that energy and matter are really two different forms of the same thing. In other words, if energy is being released (created), then matter must disappear. That's exactly what happens during nuclear fission. A large nucleus splits into two smaller ones, and a little bit of the total mass is destroyed—and lots of energy is produced. Does this mean one or two protons or neutrons get vaporized? No, the fascinating part of this is that all of the protons and neutrons lose a little bit of their mass. However remarkable it may seem, this means that protons and neutrons do not always have the same mass. A proton in a uranium nucleus weighs more than a proton in an iron nucleus. After splitting, the same total number of protons and neutrons exist; they all weigh just a bit less. This is not a recommended weight loss program for humans.

RELATED TOPICS

See also
INSIDE THE ATOM
page 18

RADIOACTIVITY
page 140

SPLITTING THE ATOM
page 142

3-SECOND BIOGRAPHIES

LISE MEITNER
1878–1968
Austrian physicist who performed the first mass/energy calculations on nuclear fission

ALBERT EINSTEIN
1879–1955
German-born physicist who provided mathematical equivalence to mass and energy

EXPERT
Jeff C. Bryan

When an atom splits through nuclear fission, some of its mass is converted into energy.

THE EFFECT OF RADIATION ON LIFE

The types of radiation discussed in this chapter are rather unusual in that they have the power to knock electrons loose from atoms and molecules. As the radiation travels through matter, it transfers some of its energy to the molecules it passes by, much like a bullet fired into a pile of pea gravel. Since electrons bind atoms together in molecules, this subatomic violence can lead to broken chemical bonds. If enough radiation damage occurs in a single cell, a large number of molecules get broken and the cell could die. If it's less damaged, it can repair itself. However, if the cell's DNA is damaged, the cell could change (or mutate) in ways that cause it to grow abnormally (because DNA directs how cells grow). These mutations and abnormal cell growth can lead to cancer. This sounds bad, especially since humans, like all living things and the planet, are naturally radioactive. Fortunately, we've evolved with rather efficient cell repair mechanisms. It appears that below a certain threshold, radiation produces no negative health effects. Some set this threshold at 100 mSv (millisieverts, a unit of dose). For reference, the global average annual dose for natural and anthropogenic radiation is 2.8 mSv.

3-SECOND BIOGRAPHIES
HERMANN J. MÜLLER
1890–1967
American biologist and winner of the 1946 Nobel Prize in Medicine who first observed changes in genes after exposure to X-rays

L. HAROLD GRAY
1905–65
English physicist and radiobiologist who pioneered studies of radiation effects on living things

EXPERT
Jeff C. Bryan

DNA can be damaged by ionizing radiation.

8 March 1879
Born in Frankfurt,
Germany

1901
Earns a doctorate in
organic chemistry at the
University of Marburg

1904–5
Works with William
Ramsay at University
College, London

1905–6
Works with Ernest
Rutherford at McGill
University in Montreal,
Canada

1907
Qualifies as a university
lecturer at the University
of Berlin

1914–18
Serves as a chemical
warfare specialist in
the German army

1918
Discovers protactinium
and nuclear isomerism
with Lise Meitner

1928
Appointed director of the
Kaiser Wilhelm Institute

1938
Observes nuclear fission,
with Fritz Strassmann

1944
Awarded the Nobel Prize
in Chemistry for the
discovery of fission

1966
Shares the Enrico Fermi
Award with Lise Meitner
and Fritz Strassmann for
the discovery of fission

28 July 1968
Dies in Göttingen, West
Germany

OTTO HAHN

Otto Hahn was often ill as a child, managing to survive both diphtheria and severe pneumonia. He never regarded himself as a good student, but both his health and grades improved dramatically in his early teens. At about the same time, he and a friend started performing simple chemical reactions with materials they could find around the house. This interest intensified when he took an evening course on the chemistry of dyes. He went on to study chemistry at the universities of Marburg and Munich.

After earning his PhD, and a year of military service, Hahn returned to Marburg for a position assisting with lectures. Hahn was interested in working in Germany's chemical industry, and learned of a position that involved international travel. His potential employer wanted Hahn to spend some time in England before they would consider him, so Hahn's advisor arranged for him to work with William Ramsay in London.

Ramsay gave Hahn a couple of radioactivity puzzles to research, and, although they were outside of Hahn's area of expertise (organic chemistry), he excelled in solving both. Ramsay was so impressed that he convinced Hahn he should pursue an academic career in nuclear science. Hahn was interested, but felt he needed more depth in this new field. After a year in London, he went to Montreal to spend a year working with Ernest Rutherford.

Ramsay then helped Hahn get a position at the University of Berlin; arriving shortly after Hahn was Austrian physicist Lise Meitner. Hahn and Meitner worked together for more than 30 years and were immensely productive in identifying many of the decay products of uranium and thorium, including the element protactinium. Meitner had to leave Germany in 1938, and shortly after that Hahn, working with Fritz Strassmann, observed nuclear fission. They didn't understand what they were seeing, so they wrote to Meitner about it. Meitner and her nephew Otto Frisch figured out it was nuclear fission. Hahn and Strassmann published their observations in January 1939, and Meitner and Frisch their interpretation a month later.

This discovery shook the scientific community, and eventually led to the Manhattan Project (the research project in the United States that produced the first nuclear weapons) as well as the development of nuclear power plants. During the Second World War, Hahn continued his work identifying the many products of fission. After the war, he discontinued his research and served as president of the Max Planck Society until 1960.

Jeff C. Bryan

NUCLEAR MEDICINE

3-SECOND NUCLEUS
Nuclear medicine uses radiopharmaceuticals to examine physiology and to treat diseases.

3-MINUTE VALENCE
As we gain a better understanding of human physiology, radiopharmaceuticals are becoming so sophisticated they can generate images and perform therapy at the cellular and molecular level. Imagine being able to kill cancer when it is so small that it can't even be located by conventional means. We may soon be able to detect and treat cancer before any outward symptoms are apparent.

Nuclear medicine involves

the injection of a radioactive material (a radiopharmaceutical) into a patient to diagnose or treat a disease. An example is ^{18}F-fluorodeoxyglucose (FDG), a radioactive sugar molecule. In our bodies, sugars tend to go to sites of metabolism. They also collect in cancerous tumors, because cancer is a sugar hog. Once the radiopharmaceutical is allowed to localize, radiation detectors can be positioned around the patient to generate three-dimensional images of the organs where the radiopharmaceutical has accumulated. The data collected from a nuclear medicine scan typically tells us more about how well the organ is functioning (physiology) than what it looks like (anatomy). By changing the chemical nature (size, shape, charge) of the radiopharmaceutical, we can obtain images from any organ in the body and determine how well it's working. Nuclear medicine scans make patients more radioactive. But, remember, that we are all naturally radioactive; nuclear medicine simply adds more that concentrates in part of the body. The radiation dose is typically low enough not to have any measurable adverse effects, and the benefits of diagnosing or treating a disease you already have outweighs any nearly negligible risk that the radiation might pose.

RELATED TOPICS
See also
CHEMISTRY COPYING NATURE
page 116

RADIOACTIVITY
page 142

3-SECOND BIOGRAPHIES
GEORGE DE HEVESY
1885–1966
Hungarian-born winner of the 1943 Nobel Prize in Chemistry who first recognized that radioactive isotopes could be used to study complex chemical processes such as metabolism

HAL ANGER
1920–2005
American electrical engineer and biophysicist who invented the cameras that are still widely used in nuclear medicine

EXPERT
Jeff C. Bryan

Radioactive substances can be used to image internal organs.

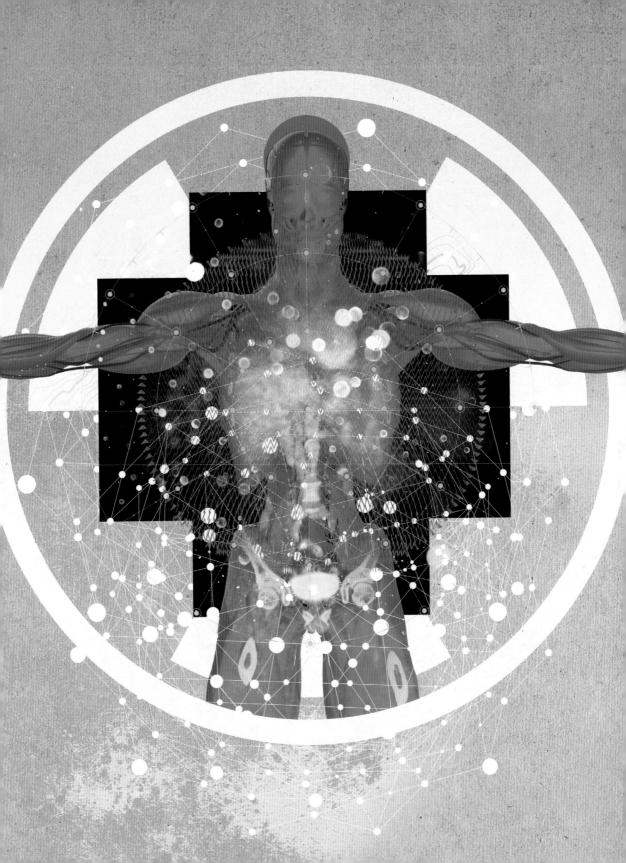

APPENDICES

RESOURCES

BOOKS

Cathedrals of Science: The Personalities and Rivalries that Made Modern Chemistry
Patrick Coffey
(Oxford University Press, 2008)

Chemistry: A Molecular Approach
Nivaldo J. Tro
(Pearson; 4th ed, 2017)

Chemistry in Focus: A Molecular View of our World
Nivaldo J. Tro
(Cengage; 6th ed, 2016)

Chemistry: Structure and Properties
Nivaldo J. Tro
(Pearson, 2014)

The Disappearing Spoon: And Other True Tales of Madness, Love, and the History of the World from the Periodic Table of the Elements
Sam Kean
(Little, Brown and Co, 2010)

The Elements: A Visual Exploration of Every Known Atom in the Universe
Theodore Gray
(Black Dog and Leventhal, 2009)

Inorganic Chemistry
Gary Miessler, Paul Fischer, and Donald Tarr
(Pearson; 5th ed, 2013)

Napoleon's Buttons: 17 Molecules that Changed History
Penny Le Couteur and Jay Burreson
(Penguin, 2004)

Organic Chemistry
Paula Yurkanis Bruice
(Pearson; 8th ed, 2016)

The Periodic Table: Its Story and Its Significance
Eric Scerri
(Oxford University Press, 2007)

Physical Chemistry
Peter Atkins and Julio de Paula
(Oxford University Press; 9th ed, 2010)

Stuff Matters
Mark Miodownik
(Houghton Mifflin Harcourt, 2014)

Uncle Tungsten: Memories of a Chemical Boyhood
Oliver Sacks
(Alfred A. Knopf, 2001)

WEBSITES

WebElements™ periodic table of elements,
Mark Winter
webelements.com

All the Nobel Prizes in Chemistry from the
official website of the Nobel Prize
nobelprize.org/nobel_prizes/chemistry/
laureates

EDITOR

Nivaldo Tro is a Professor of Chemistry at Westmont College in Santa Barbara, California. He received his BS in chemistry from Westmont College and his PhD in chemistry from Stanford University. He then went on to the University of California at Berkeley, where he did postdoctoral research on ultrafast reaction dynamics in solution. Professor Tro has authored more than twenty journal articles and has been awarded grants from the American Chemical Society, the Petroleum Research Fund, Research Corporation, and the National Science Foundation to study the dynamics of various processes occurring in thin adlayer films adsorbed on dielectric surfaces. He has been honored as Westmont's outstanding teacher of the year three times and has also received the college's outstanding researcher of the year award. Professor Tro is best known for his chemistry textbooks, which are used at more than six hundred colleges and universities around the world. About one third of all college students taking chemistry today use a textbook written by Professor Tro.

CONTRIBUTORS

Jeff C. Bryan earned an AB in chemistry from the University of California, Berkeley, and a PhD in chemistry from the University of Washington. He is a Professor in the Chemistry faculty of the University of Wisconsin–La Crosse, where he teaches nuclear and general chemistry courses, and his scholarship focuses on making nuclear science more accessible to students with limited science and math backgrounds. He has written a textbook titled *Introduction to Nuclear Science*, and coauthored a lab manual titled *Experiments in Nuclear Science*.

Stephen Contakes is Associate Professor of Chemistry at Westmont College in Santa Barbara, California, where he teaches courses in inorganic, analytical, and physical chemistry. His background is in synthetic organometallic and bioinorganic chemistry, and his research interests involve the preparation of redox-active hydrogen bonded assemblies and photoactive nanoparticle catalysts for use in pollutant remediation.

Dr. Glen E. Rodgers is a Professor Emeritus of Chemistry at Allegheny College in Meadville, Pennsylvania. Educated at Tufts University and Cornell University, he taught for five years at Muskingum College in Ohio and thirty years at Allegheny. He taught introductory chemistry on several levels, plus inorganic chemistry, and numerous interdisciplinary courses with colleagues in other faculties. He is the author of the third edition (2012) of the popular sophomore-level textbook *Descriptive Inorganic, Coordination, and Solid-State Chemistry* (Cengage Learning, International Edition 2011). Dr. Rodgers and his wife have led science travel tours around the world, taking students to key locations in the history of science.

Ali O. Sezer is a Professor of Physical Chemistry at California University of Pennsylvania. After graduating from Yildiz Technical University in Istanbul, Turkey, Sezer went to the United States to pursue an advanced degree in chemical and materials engineering. Dr. Sezer's research interests are in the area of conducting polymers, particularly nano composites of these polymers with transition metal oxides for electronics, optical, medical, bio-sensing, and corrosion protection applications. He believes in "hands-on" undergraduate teaching, especially involving students in cutting-edge research activities.

James M. Tour is a Professor of Chemistry, Professor of Computer Science, and Professor of Materials Science and NanoEngineering at Rice University in Texas. Tour's scientific research areas include nanoelectronics, graphene electronics, green carbon research for enhanced oil recovery and environmentally friendly oil and gas extraction, carbon supercapacitors, and synthesis of single-molecule nanomachines, which includes molecular motors and nanocars. He has also developed strategies for retarding chemical terrorist attacks. For pre-college education, Tour developed the *NanoKids* concept for K–12 education in nanoscale science. Tour has more than six hundred research publications and more than one hundred and twenty patents.

Dr. John B. Vincent received a BS in chemistry and mathematics from Murray State University and a PhD in chemistry from Indiana University. He was a NIH postdoctoral fellow at the University of Virginia before joining the faculty of the University of Alabama, where he is currently Professor of Chemistry. His research interests are in bioinorganic chemistry, with a particular focus on the nutritional biochemistry of chromium(III). Dr. Vincent is author or coauthor of more than one hundred and thirty peer-reviewed publications, more than fifteen book chapters, eight books, and nine patents. Dr. Vincent is currently co-editor-in-chief of *Biological Trace Element Research.*

INDEX

ACKNOWLEDGMENTS

PICTURE CREDITS
The publisher would like to thank the following for permission to reproduce copyright material:

Angelo Frei CC BY-SA 3.0 34.
Benjah-bmm 27 37, 67, 91, 93.
Bibliotheque Nationale de France 14.
Episodesn CC-BY-SA-3.0 91.
Eschenmoser 35.
Getty Images/ Bettman 130; De Agostini Picture Library 148; Keystone-France 33; ullstein bild 94.
Library of Congress 17.
NASA, ESA, CXC and the University of Potsdam, JPL-Caltech, and STScI 21.
NASA, ESA, and the Hubble Heritage Team (STScI/AURA) - ESA/ Hubble Collaboration 21.
NASA/HST/CXC/ASU/J. Hester et al. 21.
NASA and The Hubble Heritage Team (AURA/STScI) 21.
Pwnsey CC-BY-SA-3.0 91.
Shutterstock/ Africa Studio 65; Albert Russ 71; Aleksandr Bryliaev 61; Alexander Y 49; Alexey Borodin 105; AlexLMX 81; alexmak7 75; AlexRoz 35; Anastasios71 69; Anatomy Insider 151; Andrea Danti 143; Anna Kucherova 113; Antonov Roman 107; artjazz 105; Artolik 141; Arts Vector 23; Balefire 45; Bashutskyy 55; Belen Bilgic Schneider 69; Belozorova Elena 55; Bojanovic 147; Boris15 47, 125; Brenik 79, 91; Captain Yeo 75; chromatos 111, 119; chuckchee 25; chuyuss 107; CK Foto 35; cobalt88 71; Constantin-Ciprian 147; Creations 147; Crystal Eye Studio 77; Daimond Shutter 91; Dale Berman 119; dandesign86 119; Danny Smythe 65; Decade3d – anatomy online 65; Designua 17, 135; Digital Storm 111; Dim Dimich 105; Dionisvera 111; Djsash 65; dkidpix 35; Dmitry Guzhanin 63; Double Brain 29; Eivaisla 65; Ele-narts 117; ESB Professional 49; Evannovostro 97; Everett Historical 143, 145; fizkes 47; Foonia 67; Four Oaks 63; general-fmv 19, 141, 143; Georgios Kollidas 30, 50; Gergely Szolnai 67; Graeme Dawes 69; Graphic Compressor 65; graphixmania 93; gresei 113; gumbao 49; Harper 3D 97; HelenField 87; hideto999 29; horvathta 107; Hubis 143; Iaroslav Neliubov 71; Igor Petrushenko 53, 109; Igosiy 61; Ilya Akinshin 65; Iraidka 49; IreneuszB 117; J Helgason 107; Jag_cz 45, 63, 105; Jitka Volfova 127; jules2000 107; Kateryna Kon 127; Koya979 65, 141; KRAHOVNET 109, 111; KreativKolors 77; Ksander 111; ktsdesign 89; Kutlayev Dmitry 71; Login 67; lonlywolf 67; Magcom 15, 37; Makhnach_S 107; Makitalo 29; Maks Narodenko 29; Malbert 69; Mar.K 77; MaraZe 111; Mark_KA 61; martan 105; Marynchenko Oleksandr 105, 107; Masekesam 107; Mathisa 125; maximmmum 15, 29; mayakova 111; Merydolla 67; Michal Ludwiczak 71; Michal Zdu-niak 143; Michelangelus 53; Mircea Maties 29, 37; molekuul_be 43, 45, 49, 53, 79, 81, 87, 89, 91, 99, 105, 109, 113, 119, 125, 127, 129, 135; momente 69; Monet_3k 151; Morphart Creation 75, 125; Neda 37; Nerthuz 107; Nicku 114; nikkytok 35, 109; Nuttapong 69; ogwen 97; Ola-ola 89; Oleg1969 47; Olegusk 105; Olha Chernova 6; oorka 25, 33; ostill 45; Patrizia Tilly 47; Pe3K 97; petarg 135; Peter Hermes Furian 145; Peter Sobolev 119; phoelixDE 79; photoiconix 107; Photoraidz 97; pozitivo 107; r.classen 63; Raevsky Lad 99; Raimundo79 71, 117; ramcreations 47; Robert Hoetink 55; romeovip_md 63; Ron Dale 25; Rost9 8; Ruslan Grumble 53; Ruth Black 109; Sabelskaya 47; schankz 75; Sebastian Kaulitzki 135; Serg Zastavkin 55; Shawn Hempel 47, 109; Shutova Elena 111; SMSka 23; snapgalleria 133; Stephen Clarke 151; Stockphoto-graf 43; Suhecki Stanislav 147; Sunflowerr 23; tavi 63; Thailand Photos for Sale 141; Thomas Klee 113; Thorsten Schmitt 53; Tobias Arhelger 147; Tomertu 145; Tossapol 125; Tumana 27; urfin 67; Valex 99; Vasilyev 45, 55, 89; View Apart 6; Vikpit 93; Viktor88 15; VladFree 141; vovan 61; vvoe 69; wacomka 35, 99; warat42 35; Warut Chinsai 113; Willyam Bradberry 55; XONOVETS 43; yaruna 17, 43; YorkBerlin 75; zcw 53; Zilu8 45; Zvitaily 133.
Wellcome Library, London 27, 75, 117, 133.

All reasonable efforts have been made to trace copyright holders and to obtain their permission for the use of copyright material. The publisher apologizes for any errors or omissions in the list above and will gratefully incorporate any corrections in future reprints if notified.